CRACKERS IN THE GLADE

IN THE PAST.
THIS SCENE WAS BETWEEN MARCO
AND GOODLAND POINT. COULD BE SEEN EVERY
DAY - 1900's UNTILL ABOUT 1915 - BUT NEVER
NO MORE. ABOUT 300 OF THESE BIRDS ON ONE
MUD BANK FEEDING AT LOW TIDE.

PINK
SPOON BILLS
"CURLEWS"

NEVER ANY MORE
AT THIS PLACE

CRACKERS IN THE GLADE

LIFE AND TIMES IN THE OLD EVERGLADES

BY ROB STORTER

EDITED AND COMPILED BY BETTY SAVIDGE BRIGGS

FOREWORD BY PETER MATTHIESSEN

THE UNIVERSITY OF GEORGIA PRESS   ATHENS AND LONDON

The publication of this book was generously supported by the Exposition Foundation.

© 2000 by the University of Georgia Press
Foreword © 2000 by Peter Matthiessen
Athens, Georgia 30602
All rights reserved

Designed by Erin Kirk New
Set in 10.5 on 14 Berkeley Oldstyle Book
by G & S Typesetters
Printed and bound by C&C Offset

The paper in this book meets the guidelines for
permanence and durability of the Committee on
Production Guidelines for Book Longevity of the
Council on Library Resources.

Printed in Hong Kong

04   03   02   01   00   C   5   4   3   2   1

Library of Congress Cataloging in Publication Data

Storter, Rob, 1894–1987.
    Crackers in the glade : life and times in the old
Everglades / by Rob Storter; edited and compiled by
Betty Savidge Briggs; foreword by Peter Matthiessen.
        p.   cm.
    ISBN 0-8203-2066-8 (alk. paper)
    1. Storter, Rob, 1894–1987.   2. Everglades (Fla.)—
Social life and customs.   3. Fishers—Florida—
Everglades—Biography.   4. Everglades (Fla.)—Biography.
5. Fishing—Florida—Everglades—History—
20th century.   I. Briggs, Betty Savidge, 1946–
II. Title.
F317.E9S76   2000
975.9′3906′092—dc21
[B]   98-41157
        CIP

British Library Cataloging in Publication Data available

DEDICATED TO MY FAMILY.

I am especially grateful to my parents, Olivia and Bill Savidge, and Jack, my husband, for help with and support of this project

Special thanks to

Marguerite Morgan, Lem and Lucy Storter, Hub and Dolly Storter, Shelby Bender of the Quintilla Geer Bruton Archives, Plant City, Florida, and Peter Matthiessen

CONTENTS

Rob Storter, 1980.
Photo by James Cain.

I am no artist, but I like to draw things that meant so much to me. These things are in the past now, but I still remember them as though they were yesterday. They are precious memories to me. There are real stories behind all these pictures. There are so many things that were so beautiful that are gone now.

Doing what you love to do and want to do is a wonderful thing and is healthy and a blessing. I have enjoyed all this part of my past life, the fishing, hunting, and fellowshipping. There have been many dark nights that I have sat in my boat out over the water and talked to the Lord—and looked at the millions of stars dotting the black sky and at the fish glowing beneath the water.

It makes me sad to think of some of these things—and of what man has done for dollars. I think of the first few years I lived in Naples, how beautiful and peaceful—all the wonderful, natural things. All the good oysters in the bay, good clams, and so much good fishing. And good hunting so close to home. We only took what we needed to eat or make a living. Such good shelling on the beaches. So many wonderful things that are gone now. The bay has lost its natural beauty. The water is polluted. The oysters are gone; the clams are gone. The good fish are about gone. Fast boats are everywhere—the bay is so churned up and rough. The fiddler bar is gone. Man is to blame.

Robert Lee Storter, December 1968

FOREWORD

I remember with the greatest pleasure my several visits with the late Captain Rob Storter at the house in Naples that he shared with his wife, Marilea, and his daughter, Betty, who suffered from cerebral palsy (though she was never forgotten in the little house, since she was always visible through a plate glass window through which her activities and needs were constantly monitored by her loving parents). Betty died in 1986, and Rob the following year, while his wife lived until 1994, and died at home.

Captain Rob was born at Everglade in 1894. His father, R. B. "Bembery" Storter—as famously gentle as Rob himself—was, oddly enough, a very close friend of the violent and enigmatic cane planter Ed Watson, who was alleged to have killed a number of people, among them the famous outlaw woman Belle Starr. Indeed, Rob was able to provide me with perhaps the best description of "Mister Watson" I ever came across.

Rob's house was filled with his own wonderful paintings of the southwest Florida coast—untutored paintings, one might say, but "primitive" only in the narrowest sense (the way Grandma Moses paintings used to be called "primitive" but are no longer), since Rob invariably captured just what he set out to capture, and in a most fresh and lively way. That freshness is what made him a true artist.

In addition, he closely described his coastal world (often right on the painting itself), so that what he has left to us is not merely quaint or picturesque but a true historical documentation, in word and image, of a precious world and way of life that was fading very rapidly even as he recorded it. His granddaughter, the writer Betty Savidge Briggs, who in 1980 put together a smaller collection of Rob Storter's work, is to be much commended for assembling his "folk art" and journals for the present volume.

As a wanderer in southwest Florida for over fifty years, I am truly delighted to have two of Rob's wonderful wooden "books" of carved and painted fishes of the region, given me by Rob himself, and also a fine watercolor sketch, kindly presented by Betty Briggs. The painting, entitled "The Watson Cane Farm as I remembered it—1910. A Big Story of Lots of Murder at this Place in Oct. 1910," is not only well composed and colorful but also accurate enough in its detail of the plantation layout and plantings (the cane fields as well as the banana trees and coco palms and the red-flowering poinsettia tree planted on Chatham Bend in the 1880s) to have been helpful in my own research for a three-volume novel based on Watson's life and times.

Of greatest help, however, was the phenomenal memory of Rob himself, a truly delightful man brimming over with gentle stories. He was still very active when I knew him in the early 1980s, and was invariably generous with his time. He meant to see to it that his faithful record of life in southwest Florida in pioneer days was not lost to future generations, and thanks in large part to Betty Briggs's devotion to this project, it will not be.

Peter Matthiessen

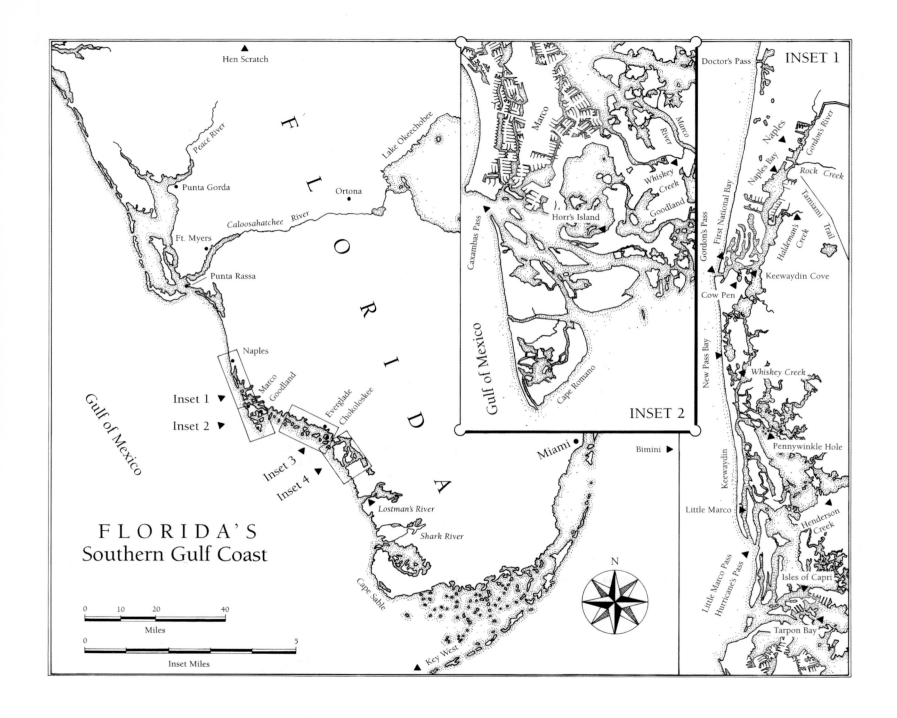

FLORIDA'S
Southern Gulf Coast

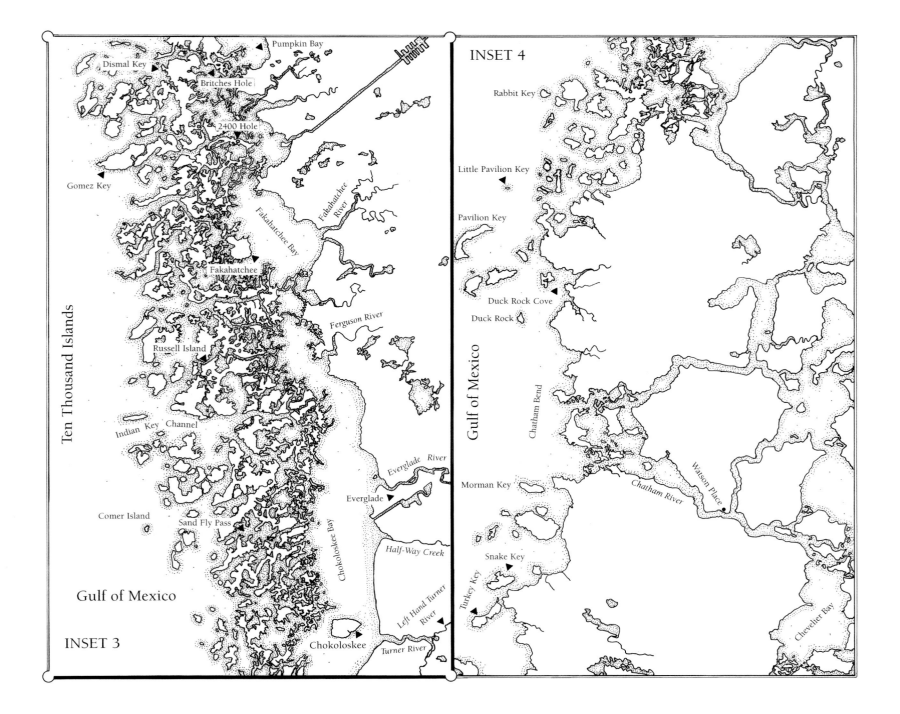

Pumpkin Bay

Dismal Key

Britches Hole

2400 Hole

Gomez Key

Fakahatchee Bay

Fakahatchee River

Fakahatchee

Ferguson River

Ten Thousand Islands

Russell Island

Indian Key Channel

Everglade River

Everglade

Comer Island

Sand Fly Pass

Chokoloskee Bay

Half-Way Creek

Gulf of Mexico

Left Hand Turner River

Chokoloskee

Turner River

INSET 3

INSET 4

Rabbit Key

Little Pavilion Key

Pavilion Key

Duck Rock Cove

Duck Rock

Gulf of Mexico

Chatham Bend

Watson Place

Morman Key

Chatham River

Snake Key

Turkey Key

Chevelier Bay

INTRODUCTION

My grandfather, Rob Storter, a fisherman, gave me his journals before he died. They are filled with stories and drawings of his ninety-three years living in the Florida Everglades. Because he was not a big man we called him "Little Papa," but most knew him as "Little Rob" or "Captain Rob," and when he was not fishing he was busy at his desk filling up notebooks or carving and painting replicas of fish. "Well, I didn't catch any fish today," he would say, "so I'm making some instead."

My grandparents Rob and Marilea Storter were born and raised in South Florida. They and their parents were Crackers: or hardworking southerners with deep ties to the land. "Cracker" is one of the oldest epithets for white southerners and has been used to describe, among other folk types, both the stereotypical poor white farmer cracking corn and the prosperous farmer who owned a slave or two, farmed and grazed cattle and hogs, and used a long whip with a tip called a cracker. This book is about both, in the sense that it covers the universal theme of surviving a harsh environment and difficult times. It is also full of fish stories. My grandfather recorded the birth of South Florida's commercial fishing industry and the seine hauls that eventually killed it. A seine haul is a big catch, the seine a large net. "Seine hauls were a joyful time," my mother said. "[They] came once a year—that was the time Daddy made the most money. It was as much a time for heartache as for joy. It was a very difficult and backbreaking way to make a living."

Mullet, for example, generally follow a pattern of moving from the inner bays to offshore waters in the Gulf of Mexico to spawn, before moving back again to the bays; the annual run is triggered by weather, and lasts from around November 20 to December 6.

As a boy, Rob watched schools of mullet pass his house on the Everglade River. It would often take an hour for them to pass, and during that time the water would turn black. After moving to Naples, he would sit for hours gulfside in a homemade skiff at Gordon's Pass or along the beach waiting for the mullet to run.

Sometimes, fishermen could only watch as the fish stayed in a narrow or shallow pass or near rocks where they could not be easily caught. These run fish would be full of red roe, fat and delicious (whereas when they returned to the bays they would be thin and tasteless). Striped or black mullet were not the only catch. Little bucks—blacks without roe—were caught before the big runs and were often used for bait. Silvers—smaller mullet with an orange band around the eyes—ran after the blacks but did not sell as well. During the summers there was a mixture of blacks and silvers.

Seine hauls usually meant no Thanksgiving dinner at home, so families would take food to the beach, often for a fish fry, and while someone watched for the first sign of the run, everyone would eat. Pelicans diving, porpoises jumping, or mullet skipping through the water would send the men scrambling to the boats. After the net was cast, families would help pull the ends of the net up on the beach. Word spread quickly, and sometimes tourists would come and watch, and as the fish were pulled closer to the beach, they would help themselves. "He never minded sharing a few fish. After all, there were thousands of pounds," my mother said.

Not every fisherman owned a big boat or seine net. Sometimes, after the fish were pocketed, or trapped in the net, Rob would invite "little fishermen" to strike inside the seine with a rowboat and gill net, a net set upright in the water to catch fish by entangling their gills in its mesh. Then the net was taken to deep water, and the fish dipped into the boats, a long arduous task. Even in the cold, sweat could be seen dripping off the fishermen's noses.

I remember the laughter, sunshine, and good food, but it was only after the fish were taken to the fishhouse, unloaded onto a conveyor, shoveled into heavy steel baskets, weighed, and the men paid that the seine haul ended. Uncle Lem remembers being released from school during run season and working day and night so forty thousand pounds of netted fish wouldn't spoil. When the season closed, they had five days to get rid of the fish, and since local fishhouses had a surplus and wouldn't take any more, they headed to a Fulton County, Georgia, fish market, "with a driver named Blackie who kept running off the road and taking No Doz."

The men fished for as little as one cent a pound and as much as twenty-five cents. One year the prized roe, shipped to Japan, sold for thirty-five cents a pound. During World War II guiding jobs were scarce and there was a limit on how many fish you could catch. Many fishermen smuggled fish to Georgia. One man brought his catch to an isolated shore and loaded it onto a hearse, hiding the fish under the seats, then proceeded on to Georgia. My mother remembers earning seventy-five dollars helping with one haul right after the war—much-needed money, as jobs were scarce. My aunt bought a twelve-gauge shotgun with money earned from another haul. The kids never made money, but the fun was pay enough. If we were lucky and the seine occurred near a winter visitor's unoccupied beach house, our parents would let us go up into the yard and pick seagrapes, which we ate until our tongues were raw and purple.

Everyone depended on run season and took turns—Rob might get the haul one day, someone else the next. But greed, exhaustion, and sad times were part of the package as well. Once a hungry fisherman threatened others with a gun to keep them away from the water where the fish were running. After a fire destroyed the top of the tent Rob and Marilea lived in, Rob came in so tired from fishing all night that he didn't even notice.

My dad, Bill Savidge, often fished with Rob and remembers the 1946 run season, waiting long hours in the Gulf for the run to start. Just as the school got through Gordon's Pass they started disappearing, something Rob had never seen happen before. Suddenly, fish started floating to the surface. A patch of

Mending a net in the backyard by the net shed. Photo by John Briggs.

unpredictable red tide had wiped the school out—thousands of them, just like that.

Seine hauls are now illegal. Rob would not have welcomed the ban, but neither would he have fought it. In 1910 when he started fishing for a living there were no catch restrictions, licenses, or designated seasons. No one had to tell the fishermen when to close the season; they regulated the catch themselves. The earliest nets were made of cotton and had to be pulled out, "limed," and dried after every catch. Then came linen and, eventually, nylon or monofilament, followed by the king of all nets, the seine, which enabled the men to catch as much as a hundred thousand pounds in one strike. One of my early memories is of my grandfather standing over a net mending holes that fish had made the night before.

The Storters are the founding family of the town of Everglade (later changed to Everglades City). Rob's grandfather, George Washington Storter, who as a young boy saw Napoleon riding on a white stallion, emigrated from Alsace, France, in 1835, with his parents and brothers. After the death of his parents, George, then seven years old, was adopted by the clerk of the circuit court in New Orleans, Joseph Genios, a Frenchman who became a judge. The judge is said to have used him as a jockey in plantation races because of his small stature. One of George's brothers was adopted into a wealthy family, and George remembered seeing this brother sitting on the steps of his new home as he walked by. His brother stuck his tongue out, and George ran. The brothers never saw each other again.

Thanks to a suggestion from an old man in Demopolis, Marengo County, Alabama, I discovered two journals in the Alabama State Archives recording George's business dealings from 1856 to 1866. He and James Cunningham of Eutaw, Alabama, owned and operated tin shops in Demopolis and Eutaw. The journal records such details as the purchase of a Negro girl for forty dollars from the town doctor around the time George's second child was born. An 1860 Marengo County census lists George as owner of a residential hotel with thirteen lodgers, all men, including a Bavarian shoemaker, a Swiss confectioner, French chicken husker and a deaf and mute Bavarian photographer. George's wife, née Harriet Elizabeth Hallbrook (Betty), a plantation owner, was the only female residing at the hotel.

Civil War pay vouchers reveal that George served as a sergeant throughout the Civil War in the Second Alabama Cavalry and was paid forty cents a day extra for bringing along his own horse. A slave from his wife's plantation accompanied him. The company made a brief raid into Florida; it is believed that this was George's first glimpse of the state. Toward the end of the war George served as an escort to President Jefferson Davis, and upon his discharge he was awarded a pension of one hundred and twenty dollars. He refused to send in his pay vouchers, however, and so did not collect the pension for five years. The claims process would have put him under oath—and George balked at this, for he had religious scruples against "swearing"!

Within twelve years after the war's end he was widowed. Around 1875, following the death of Willie, his firstborn, at age fifteen, George left in a wagon with his two surviving sons, George Jr. and Robert Bembery (R. B.), for Pine Level, Florida. They camped along Horse Creek, near Pine Level and Fort Ogden. Pine Level, the Manatee County seat, had been turned into open range owing to reckless lumbering. George stayed four years.

It was said of Pine Level that life and property were not safe. According to one old-timer who knew him, George had been in town one night when he heard a man say, "Well, we generally kill a man here for Christmas." George did not pay the remark any mind, but on coming to town the next day he found a local rogue, Bill Key, dead in the road. Bill was known to have abused a little man named Jim Driver, who in turn had threatened to kill him.

Father and sons settled in the Everglades in the late 1800s and grew sugarcane on the same land for close to twenty years without replanting or fertilizing after the start-up year. R. B. carried the mail on the schooner *Bertie Lee* from Everglade to Key West and Tampa. George Jr.—whom Florida historian Charlton Tebeau credits as the founder of Everglade City—became the town's first judge, ran a successful general store, and built a large home, which Barron Collier eventually turned into the renowned Rod and Gun Club, an inn for winter visitors. A traveling dentist recalled spending the night at George Jr.'s home and sleeping in a room with fifteen children.

Everglade, though not known for its violence like Pine Level, was hardly a peaceful paradise. The remoteness of the Ten Thousand Islands made an ideal refuge for transients and renegades, and law enforcement was virtually nonexistent. The nearest peace officers were more than a day away, in Key West, Fort Myers or Tampa, and those areas could not be bothered with the scattered population of the Everglades. Local citizens therefore made and enforced their own laws, which often as not were little more than mutual agreements. Cohabitation frequently was enough to signal a common-law or informal marriage. As for divorce, it was uncommon; separation by agreement was by far the more usual arrangement.

This frontier unruliness occasionally made for some excitement. In 1915, when one of the Homestead bank robbers, Leland Rice, was shot dead for the reward money right outside my grandfather's door, my grandfather helped bury him. Sadly, some time later two of my grandfather's brothers were arrested for smuggling aliens into the country from Cuba. The infamous and enigmatic Ed Watson ate many meals with the Storters when he came to town for his mail. The family declined to speak badly of him, even though he was accused of killing his farm workers after the body of one rose to the surface of the water. It is probable that Watson murdered other farm hands. He was killed by a mob of Everglade men before he could be tried in court. In spite of Watson's dark side, Nannie Storter, George Jr.'s wife, was like a second mother to the Watson children, and as a youngster my grandfather became good friends with Ed's son, "Colonel" Lucius Watson, who was like another brother to him.

Everglade was home to the Storters for many years. The family moved to the coastal towns of Naples and Fort Myers after Barron Collier bought Everglade in the early 1920s.

Rob's journals are first-hand documents of Florida's last frontier—a wilderness that was as dangerous as it was colorful, where you were as likely to meet an itinerant preacher or a renegade hiding from the law as a scientist studying cormorants. He wrote most of the entries in the last thirty years of his life, wanting to preserve his memories but aware that people did not have time to sit down for long and listen to storytelling. He spent many soli-

Rob Storter as he was best remembered in his last years—working at his desk or the kitchen table, creating art work and writing down his memories.

tary and peaceful hours after coming in from fishing, gardening, or skiff-building, recording his experiences while Tiney, his little black dog, sat on his lap. In the last years before his death, Tiney was his constant fishing companion.

In the late 1960s, my husband, an artist, saw a model boat Rob had made out of found materials and recognized his artistic bent. He gave Rob a watercolor book, paints, and brushes, which started him drawing. As is true of many "outsider" or folk artists, though, artwork took shape not only on paper, but also on shirt-packaging cardboard, concrete, and scraps of wood. The net-shed door became a record of hurricanes and tides, corks were transformed into fishermen in wooden skiffs, and wood scraps took the shape of brightly colored fish and birds in wood boxes.

When a journal was filled, he would keep it around for a time and share it with family and friends. Every so often I would ask to have them. From time to time, too, Marilea would gather up a stack of drawings and sell them for a nickel at garage sales. His memories and his stories are the substance of these writings, these drawings. I did not want them to be lost.

After Rob started his journals, I encouraged Marilea to write. My mother did the same with her grandfather, William Summerall, who up until shortly before his death at age 103 walked twelve miles a day to pick up groceries. Marilea's words are, at times, haunting as she tells of moving across central Florida, living out of a wagon while her father cleared land for a living, and of the hardships women suffered. Fifty-eight years of caring for a daughter with crippling cerebral palsy left an indelible mark as well. Although the bulk of this volume is devoted to Rob's memories and stories, occasionally Marilea's reflections express a point perfectly. It seemed fitting to present their stories together, since they shared their lives so thoroughly and for so long.

Rob was born at home in Everglade during a 1894 hurricane and lived there until 1916, when he moved to an oyster bar in the lower end of Naples Bay with his new bride, Cassie, who died a short time later, most likely of typhoid. Everglade was becoming a busy county seat, a fact that did not entice him to stay.

From an early age his dream was to be a "graduate fisherman"; the census of 1910, when he was just fifteen, lists him already as following that occupation. Soon after Cassie's death he married Marilea Summerall; meanwhile, Naples quickly became home, and in the early years they knew every local family and winter visitor. Marilea raised four children— Lemuel, Marguerite, Olivia, and Betty—in tents, houseboats, and their first real home on Pulling (Goodlette) Road. Between fishing seasons Rob guided, making six dollars a day. He also built skiffs (though he never considered himself the carpenter his brother Wilbur was), and when necessary he would do odd jobs to support the family. During World War II, like many other fishermen, he helped build the army base in Fort Myers. Although times were not always easy, he managed to make a comfortable living from fishing and guiding his whole life long.

In the summer of 1980 I took my grandfather to Everglades City, and he was surprised at how ghostlike the town seemed. We tried to find his old homeplace, finally settling on a site occupied by a new house on stilts. "All the trees are gone," he said; "we ate fruit every day off those trees." When he recognized a lone palm, his face brightened. We went to Smallwood's General Store on Chokoloskee Island, where he embraced Thelma, the surviving daughter of Ted Smallwood, who moved to Chokoloskee in 1897 and became owner of most of the island and was postmaster for twenty-five years. (Thelma has since died, but the store has been preserved as a historic landmark.) Two strangers walked over to us, and my grandfather pointed to a rotting wooden marker in Chokoloskee Bay and told them there was a "big story" behind that marker. To Rob, any story he had to tell was a "big story." They walked away as he started recounting, turning their attention to a rack of postcards for sale on a counter—the counter

Rob with one of his model ships, the *Nanna*.

Rob was rarely without his long-brimmed khaki cap.
Photo by John Briggs.

Building a skiff under the net shed. Photo by John Briggs.

"Old Moge," the town character who was always trying to catch Thelma Smallwood's eye, had slept on when my grandfather was a boy. As though the memory was enough, his eyes smiled at the water and the decaying marker. We stopped and visited friends, Reverend Hancock and his wife. The old Church of God preacher, living in a trailer at Chokoloskee, said he spent most of his days in bed. Rob—eighty-five years old—proudly said that he was still fishing, and they tearfully embraced. I picked up a T-shirt before leaving town that day: AMERICA'S OLDEST INDUSTRY SINCE 1602—ENDANGERED SPECIES, THE COMMERCIAL FISHERMAN. On our way out of town Rob took a long look at the sign on a street corner next to the Rod and Gun Club—the house that had been his uncle George's. It read Storter Avenue. "They can have it," he said. "I got my good out of her."

My grandparents lived in a time when a man might sell a land claim for three hundred bunches of bananas and the New York fashion industry would pay five dollars for one Deep Lake plume feather. Growing up, I heard stories of panthers, alligators, buried gold, yellow-dog sausage, and trips through Hell's Gate and the Cow Pen. I raced naked with my sisters and cousins on pristine beaches, chased King Henry fiddler crabs on sandbars, combed the shoreline for purple fan-tailed pectens, and enjoyed Swamp Buggy parades and picnics of fried mullet, prized red guavas,

Rob's net shed, where he kept a record of hurricanes, fish he had caught, and other interesting facts, writing directly on the door and side of the shed.

A hand-carved knife, made of a rib bone.

Fish plaque made by Rob.

The inside of one of Rob's fish boxes.

and seagrapes. During a strike, or catch, we pretended to be Indians and stomped on the bottom of the skiff to scare fish into the nets. I experienced the thrill of hooking a tarpon, only to have my grandmother grab the pole, thinking I was about to be pulled out of the boat.

As my grandparents neared the end of their long lives, darker stories emerged: the preacher who exposed himself to my grandfather's eight-year-old sister Winnie as he walked the youngster home from cleaning the church; the respected Chokoloskee family man who molested his daughter; women worn down from excessive child bearing, sometimes the result of incest, who used clothes hangers to abort pregnancies; friends who died of untreated syphilis; and fishermen, no longer able to make a living from fishing and guiding, who turned to the drug trade. Yet my grandparents followed a simple philosophy in a not-so-innocent world: "It is God or nothing." That's what Rob said when a steel basket full of mullet fell on his brother's head, almost killing him, and he believed it all his days. Until a few months before he died, my grandfather went out fishing regularly with his skiff and net—backbreaking work even for someone much younger. When a strike brought only catfish that even the pelicans wouldn't touch, he did not complain. He loved fishing, pure and simple. "I never would have done it for so long if I hadn't loved it," he said.

While there is no doubt that religion played a key role in his life, I believe my grandfather's love for fishing was his special gift. "It is a blessing when a man can do what he loves," he would often say. When he was no longer able to put in long days and nights out on the water, he spent more time at his desk, whittling and painting replicas of the fish he had caught all his life. As his eyesight failed, he began painting the fish in pale colors and putting them to rest in handmade covered wood boxes. In the last hours of his life he would raise his arms in a gesture that my mother said was a sign for God to take him home. He awakened shortly before he died, smiled, and whispered that he had just dreamed he was in a bright, sunny room full of candy. My grandmother, in

Rob carved many boats like this guide launch.

contrast, did not wait so patiently and sweetly at death's door. In her final moments I could hear her saying, "If you think you want me, come and get me." While my grandparents were very different in how they saw the world, they were both strong-willed and fiercely loyal to each other and their handicapped daughter (who died a year before my grandfather).

Although the journals are not presented in their entirety, they do appear in chronological order within each section and have been carefully selected to reveal various aspects of Rob's life. While some of the writings were recorded in bound journals, others, written on dozens of looseleaf sheets of all sizes, had no particular format. The task of organizing by date—as best I could—and deciphering handwriting took two years. I have tried to keep my grandparents' words and spirits intact, editing only for clarity and easier reading. Notes are included to give background information. I accept full responsibility for any inaccuracies or mistakes (theirs or mine).

When I finished compiling these journals I showed a few to the editor of *Sports Afield* magazine, who was eager to share them with a friend of his. That friend turned out to be Peter Matthiessen. Ironically, my grandfather had talked about a "gentle, kind man from New York who sometimes visited and listened to his stories." The family was unaware that Mr. Matthiessen was that same man; he had been gathering information for his book

Killing Mr. Watson, in which my grandfather was the character Hoad Storter.

In a paradoxical way, the end of seine hauls and other practices my grandparents took for granted is cause for celebration. The bird population is now beginning to come back; fish are again plentiful (although signs warning of mercury poisoning are still posted in areas of the Everglades); and Florida has moved forward with a major restoration project for the Glades. Yet it is anyone's guess what will happen to the "River of Grass" as politics and nature continue to follow their respective courses.

One thing I know: I will never forget Naples Beach, pulling on the net with my sisters and cousins, under my grandfather's loving eye. I thank my parents for their willingness to share us with our grandparents by living next door to them—something that, sadly, is not possible today for most families. I will never forget my grandfather sitting in the bow of his skiff, steering us through the darkness, toward home, and the calm we felt even when the water was rough and the sky threatening. That an era has ended and an estimated fifteen thousand fishermen have lost their livelihood is sad, but the attempt to restore nature's balance is important, and has been long in coming. "The fish are disappearing, just like the land," my grandfather said. He wisely knew that his experiences were for his time and place alone, that they were unique—and, I believe, that is why he recorded them.

Inside of one of Rob's wooden "fish books" that he whittled and painted.

Fish book cover.

My grandparents' net was always full, though their lives were weighted by experiences I'm not sure many could endure with a smile or humor. Now they are gone, and their spirits are with me. For that I am forever grateful.

Many have contributed to this book. In addition to my grandparents, others include my father, Bill Savidge, who gives a poignant account of Rob's last fishing trip, and George Storter, who left accounts of the Watson killing, bootlegging, and alien smuggling.

Family and friends shared my grandfather's love for fishing. Boyhood friends Nelson Noble, Lucius Watson, Clarence, Frank, and John Brown, Bill Stephens, and Bill Gardner were like brothers. His fishing and guiding years were spent with those he fellowshipped with; among them were Forrest Walker, Ed and Joe Townsend, Jack and Cap Daniels, and Mack and Charley Johnson. Family held a special place throughout his life. He worshipped and fished with his brothers, George, Hub, Wes, Claude, and Wilbur, his four children, a special nephew, Bem, his twelve grandchildren, and numerous great-grandchildren. People mentioned by name in his journals and not otherwise identified, either in a note or in brackets in the text, are all neighbors and friends of my grandfather's.

Rob, holding one of his little dogs; Betty; and Marilea, holding her great-grandchild, Jena Briggs. Photo by John Briggs.

CRACKERS IN THE GLADE

Within the drawing:

1914

AS I REMEMBER IT.

(DAD ROWE) RETURNING
HOME AFTER CHURCH
AT CHOKOLOSKEE.
11s HALL 1914 —

DAD WAS A TRAPPER
AND FISHERMAN —
A GOOD OLD CHRISTIAN
MAN — HE FISHED
MOSTLY FOR TROUT
WITH BAMBOO POLE (36

Dad Rowe (1914), trapper, fisherman, and Christian.

ONE 🐟 EARLY DAYS IN THE EVERGLADES

"UNCLE GEORGE BOUGHT THE TOWN OF EVERGLADE FOR $800." ROB

I played hide-and-seek under the great madeira [mahogany] tree in front of Uncle George's house. Uncle George had a general store at his place where I remember seeing him measure alligator hides all day. One of Chokoloskee's earliest settlers, Mr. C. G. Mc-Kinney, died one day of a heart attack while visiting Uncle George's store. I was making a coon skiff that day.[1]

Grandpa's tin shop was on the riverbank [of the Allen River], and he lived in the upper part of it until he died at age ninety-two. Near the end of his life he tended his garden near the tin shop, often taking refuge in a shack in the middle of the garden to rest on a cot and read. At times he would tie some of his possessions up in a red handkerchief and swing it over his shoulder, telling us he was going back to Alabama. He'd walk down the road a bit, turn around, and go back. He had a stroke in 1921 and was bedridden for a short time before dying in my uncle George and aunt Nannie's bedroom just before his ninety-third birthday.

Uncle George's house was the first house to be painted in Everglade and eventually was sold to Barron Collier and became the Rod and Gun Club. The first school began in Uncle George's house in a bedroom. The students, about twelve of them, were taught by J. W. Todd.[2] My dad brought Mr. Todd to Everglade from Key West on his schooner. There was always a full house at Uncle George's.

My uncle, founder of the town of Everglade, was justice of the peace and later the first judge of Collier County. He married people and settled lots of disputes. It was at this time that the name Chokoloskee was submitted to the Post Office Department for approval for a post office at Allen's River, but the name was rejected. My dad suggested Everglade, and it was approved. It was not until 1923 that an *s* was added and Everglades became the county seat.[3]

My grandfather George Washington was an immigrant from Germany in 1835. His parents died of yellow fever after they arrived in New Orleans. He and a brother were adopted into families. One brother took the new family name and was never heard from again. At twenty-eight, my grandfather married Betty Hallbrook of Eutaw, Alabama, and they had three other children, William, Mary Lula, and Jimmie, all of whom died. Only my uncle and dad survived.

I was told that asthma and a desire to move to a warmer climate were why he [George Storter] moved to Florida. He had $200 in silver and gold saved up, and since there was little money circulating, he was considered rich. On a June morning in 1877, with two sons, he started out in a covered wagon to Fort Winder, Florida.[4] They crossed the Peace [Peas, then Pease] River at Fort Ogden. They stopped along the way to help farmers in exchange for potatoes or vegetables or a few meals, and when they finally arrived in Fort Winder on July 15 they liked it. But my grandfather did not stay in Fort Winder long. He traveled southward and settled briefly on Horse Creek, then returned to the Fort Winder/Pine Level area and bought a place from John Cash [a West Indian sailor] in 1878. A year later he purchased a place from Gus Lawrence in Pine Level.

In September 1881 my grandfather and a neighbor went to Everglade, then called Chokoloskee. He traded his single ox-cart for a sloop made from a yawl and hired John

Waterfront view of George's home and trading post, 1917. Photo courtesy of Florida State Archives. Note the scow (large flat-bottomed boat with square ends) in the foreground.

Standing in front of the general store (*left to right*): unidentified man, Nell and Dot (two of George Jr.'s daughters), George Jr., George Sr., and Robert Bembery. Photo ca. 1910, courtesy of the Florida State Archives.

Rear view of George W. Storter home, 1915. Photo by Fenton Garnett Davis Avant, courtesy of the Florida State Archives.

Cash to sail it from Charlotte Harbor to Everglade. They stopped nights on the trip down and on the way back at the only inhabited places along the way—Punta Rassa, where a telegraph office was, and Marco, where Captain W. T. Collier lived, and Wiggins Pass, where Joe Wiggins had a small trading post and apiary.

In Everglade they found Madison Weeks and his family living along the river.[5] The family raised cabbages and turnips for shipment to Key West. My grandfather, uncle, and dad found the Weeks family living in a small house that consisted of two twelve-foot rooms with a passageway between the two. The house was made of palmetto logs with cabbage palm fronds for a roof. Also living along the river was William Allen, a widower at the time he settled in Everglade. Mr. Allen owned and occupied all of the townsite from 1873 to 1889, prior to which time a plume hunter with five daughters, William Clay, had the property.[6] Allen built a house on the site that my uncle later bought.

My dad left a detailed account of the early settlers in Everglade at the turn of the century. He wrote that Allen planted his crops along the riverfront, from the mouth to Dupont on the east bank. My grandfather took advantage of his new friendship with Allen and went to work harvesting Allen's crops. The first crop

Uncle George's home, the first house to be painted in this part of the country. Grandpa's tin shop is shown on the riverbank.

Rod and Gun club; formerly Uncle George's home.

Betty and George Sr. before they married. Photo from a tintype, courtesy of Nan Alderman Dean.

operated the schooner that took the produce to Key West.

When William Allen died, Uncle George bought the Allen property for $800. The town was unsurveyed, but Allen had been given a certificate that gave him rights to purchase. My uncle's property included all the present Everglades townsite. He began cutting buttonwood by the cord to supply the Key West market.[7] My grandfather, a tinsmith, got tin from Tampa and made cans to hold syrup that my uncle and dad made from the plentiful cane crop in the area. One acre made about seven hundred gallons of dark syrup. My family canned between 200,000 and 300,000 gallons each year. Their crop was grown on Half-Way Creek, so called because it was halfway between Everglade and Turner River. They brought it to Everglade by barge, where it was

processed. Both my uncle and dad processed the cane, each by his own method. My uncle used iron kettles in which he cooked the cane juice, and my dad used an evaporator-type pan. My dad, also called Captain Storter, reported that he grew cane on the same land for close to twenty years without replanting or fertilizing after the first year.[8]

My grandfather made the first fishing spoon to lure snook. Mr. Wilson had the first rod and reel we had ever seen.[9] My grandfather made this spoon in his tin shop down by the river. We had always used pork rinds, feathers, buck tails, or white rags. But the fish went crazy over that shiny tin spoon. The next year Fluger came out with a spoon like Wilson's, only the hook was fastened solid. Later in his life, my grandfather made gas tanks for small motor boats.

consisted of tomatoes, cucumbers, and eggplants. All produce was shipped by way of schooner from Everglade to Key West, where it was transferred to a Mallory Line steamer bound for New York. Mr. Allen's son, William,

"HE TOLD OF AN OLD MAN THAT PICKED UP A BABY THAT WAS HIDDEN UNDER AN INDIAN WOMAN'S SHAWL AND THREW IT UP AGAINST A TREE AND KILLED IT." MARILEA

When the Storters first went to Everglade they crossed the Peace River at Fort Ogden [south of Arcadia], where my grandpa David and Papa [William] lived and ran a little ferry. My granddaddy had lived with his brother on a cotton farm, and when he left he was given two slaves to take with him. My papa remembered the day the Storters crossed the river with their mule wagon. They charged the Storters fifty cents to cross. He said that they

were the most stubborn mules he ever saw and it almost tore up their little barge and they sank and had to unhitch and swim. They charged a horse and wagon fifty cents and a horse and buggy twenty-five cents, horse and a man twenty cents and one man ten cents.

My daddy said George Storter crossed the Peace River twice a week to hunt cattle at Hickory Bluff [near Charlotte Harbor]. George would ride a mule. My granddaddy let him cross free. Mules were really scared of water.

The old ferry leaked so bad, every morning it would sink. Grandpa Summerall took cattle from Hickory Bluff to George Storter for him to keep 'cause the Negroes were stealing them.

My granddaddy sat at night and talked about the Indian War [the Seminole War of 1835]. He went on scouting trips—hunting for Indians in the Okefenokee Swamp [on the border of Georgia and Florida]. They would tie the horses up without water or grain for three days at a time. They lived on hardtack [biscuits] and water, and they cut poles and slept on them. The gators splashed water on them at night while they slept. One night two drunk men came along and said, "Old man, get up, we want to hear you talk about the war." He told of an old man that picked up a baby that was hidden under an Indian woman's shawl and threw it up against a tree and killed it.[10] Indians burned a house nearby and killed people and burned the barn and crops. He went there and found a young woman who was still alive and two little boys. One boy was chopped in the head with a hatchet, and the breasts had been cut off the young woman.

My granddaddy was in two wars, the Indian War and the Old Rebel War [Civil War], and lived to draw a pension in gold. My grandmother Elizabeth Douglas was young when she married and was from Hillsborough County, Florida. My daddy was delivered by his daddy. My mother originally came from Georgia. Her first husband had left her with nothing and she married my papa. My aunt told me they got me out of the pond near the house. I attended school at Zephyrhills, but left when I was in the second grade. I never attended a full term in one school. Sixth grade was the last of my schooling. Papa was cutting poles for the electric company in Tampa then. When we lived in Bloomingdale [near Tampa] somebody shot in our house and peppered one wall during the night. Papa jumped up to

Doad and Pat along the shoreline (1904).

get his gun, but Mama had washed his hunting jacket and he couldn't get his shells quick enough, so whoever it was got away. We never knew the reason for the shooting. A bloodhound traced the tracks to a houseboat not far from us, but Papa would not let the sheriff press charges. We lived in a small shack a short time in Crystal Springs [between Zephyrhills and Plant City]. While we were there the children and Mama came down with chills and fever. Papa gave us a chill tonic of senna-leaf tea and castor oil. When we were sick Papa went over to Zephyrhills and came back with some loaves of bread and apple butter—the first loaf of bread I had ever seen. I was about seven years old.

"MR. KINGSTON PAID US A DOLLAR FIFTY FOR A DAY'S WORK." ROB

When my cousins were of high school age my uncle George bought the *Terre Haute* so my dad could operate it and take his children to school in Fort Myers. I went along sometimes and relieved Papa at the wheel, but I didn't go to school anymore. I completed the sixth grade. I remember my first day of school. It was in the parsonage and the teacher's name was Gant. I ran around the house several times before he caught me. He talked to me, gave me a nickel, and had no more trouble from me. There were about twelve of us in the school, mostly Storter children, except four were the teacher's children. We did not wear shoes to school, only occasionally when we went to church.

I loved operating the *Terre Haute,* a fine forty-foot launch equipped with a sixteen-horsepower standard motor. My brothers and I worked on it, scraping, sanding, and varnishing it, before my uncle bought it from Mr. Kingston, who invented the Kingston carburetor and was a winter visitor to the area. Mr. Kingston paid us a dollar fifty for a day's work.

Mr. Collier hired my papa to take his New York friends out on the *Terre Haute.* After a few trips of drinking and girlie sprees, my papa said he could not be a party to that way of life.

Because there were no roads to Fort Myers from Everglade, the *Terre Haute* took passengers there. I remember lots of things about Fort Myers.[11] There were no cars, but lots of horses and buggies on Main Street. You could rent horses and buggies at the livery stable. There was a bar, and you could smell it before you got near it; two grocery stores, R. A. Henderson and Heitman; Evans Franklin Hardware; and a fruit stand called Dancy Stand where you could buy candy, fruit, peanuts, tobacco, and other things. It was located at the foot of Ireland's Dock. There were three docks—the railroad dock; Ireland's, where good old Gulf gas was sold wholesale; and at the head of the dock on the north side was a marine ways [dry dock]. There was the City Dock, where the Kinzey Brothers unloaded shell that was hauled from over near Punta Rassa or Shell Point.[12] The Kinzey Brothers had mail routes up the river to Alva and La-Belle and then south to Punta Rassa and up near Charlotte Harbor. A little steamer, the *Gladdys,* up the river, was a steam stern-wheeler, and at the City Dock was a fish retail market for fresh fish and oysters. There was also a furniture store. You could ride all over town in a few minutes on a bicycle. I had one, and the front wheel was warped, so I had to carry a little can of lard to keep the wheel from rubbing too hard. In those early days I do not remember seeing a single car, but later I remember seeing some Model T Fords. Sikes and Hill had a garage on Main Street on the west side with a filling station with hand-cranking pumps. You pumped the amount of gas into a big glass and then let it run into the car tank.

I remember Mr. Collier's hotel and Annie Stephens's boardinghouse at Marco. Later, there was a fish house where a gas dock was. There were more houses farther back. The main street was a gathering place for loafers. There was only one store and post office. Later, clams were canned at Marco. Everybody had a job, some working in the factory, some building boats and repairing, some fishing. There was a boatways there where you could pull out a good size schooner. My dad pulled his schooner out there for repairs. They built several schooners there. In sailboat days Mr. Collier took pride in building these boats. Later Jim Daniels's sons, Henry and Frank, took over the boatbuilding and did a good job of it. They moved to Fort Myers and went to it in a big way. In 1916, we had to get our groceries at the Marco store as there was no store in Naples. I can remember there would always be a crowd of grown, barefooted boys in front of the store playing marbles or boxing. Later, the clam business moved to Naples until the clams played out and never came back.

It would be hard to believe that there was once a railroad into Marco (for what, I don't know).[13] It didn't pay and soon was taken up. All the pilings for the bridge, crossing the Marco River, were pulled up and the rails moved. At one time the old country road that went into Marco ended at where Isles of Capri is now, and a flat barge ferry carried the cars across the channel to Marco. While waiting for the ferryboat to come over, sometimes thirty minutes to an hour or more, the sand flies and mosquitoes would almost eat you up. The barge could only carry about two small cars at a time, so you had to wait your turn. And what a relief when you got out in the channel and a little breeze blew the mosquitoes and sand flies off.

Downtown Old Marco at the bay front (1909).

I am sure there is plenty of gold that has never been found or never will be. I have the story of three brothers that found and hid treasure up the Peace River. After the treasure was buried, one brother went back to Spain to get a boat while the other two stayed to watch out for the treasure. They waited long enough to raise an orange grove. They waited forty years for the brother to return, but one of these two waiting brothers died, and the other was getting old. One day he told an old man that was living with him to go with him and he would show him something. They went down to the river and, standing on the bank, he told this man that his brother had been gone forty years and he knew he would never return and just right over there was more money than he could ever spend. This was up the Peace River in the bend of the river, and when he said "just over there" and pointed it was just a short distance to an island with large cypress trees on it. This was near where his orange grove was. There are a few of the old orange trees left there. These brothers were Spanish and their name was Gonzales. The old man that Gonzales told all this to was Roberts. Roberts told this story to a man that lived in Fort Ogden named Heiss, and I talked with Mr. Heiss. He told me the whole story. He said he had spent fifteen years searching for that treasure, but never found it. Mr. Heiss had a chart. He is dead now. He said when he was a boy he would go down to the river often and swim for several hours. They would climb up on the bow of the old schooner and dive off. The old boat finally rotted in the river. He said he had no account of how or when the boat was put there, only he heard this was a pirate boat, and the whole crew, after burying the treasure, got in a free-for-all fight and all were killed.

Nannie Waterson Storter, George Jr.'s wife, is wearing a white dress with dark bow. George and Nannie's house is in the background. Photo ca. 1910, courtesy of the Florida State Archives.

George Storter Jr. pulling an Everglades skiff, 1906. Photo courtesy of the Florida State Archives.

"GATEWOOD"

A GOOD METHODIST PREACHER.
SURELY HE MUST OF BEEN A CHRISTIAN
AND HAD SOME LOVE. I ONLY REMEMBER
HIM AS A GOOD MAN THAT USE TO COME
TO EVERGLAD WHEN I WAS A SMAL BOY
AND WHAT I WAS TOLD ABOUT HIM.
HE LIVED IN PUNTAGORDA. BUT WOULD
RIDE A HORSE TO NAPLES AND
BORROW A ROWBOAT AND ROW TO
EVERGLADE TO PREACH. THERE SURE
WASENT MUCH MONEY THOSE DAYS.
HE MUST OF HAD SOMETHING, THAT'S
LACKING THESE DAYS.
I HAVE HEARD MY PARENTS MENTION
HIM MANY TIMES, AND IT WAS ALWAYS
SOMETHING GOOD. HE WOULD LEAVE
HIS HORSE WITH CAPT. CHARLIE STEWART
IN NAPLES, HE ALSO MINISTERED TO
THEM. COULD YOU AMAGINE A PREACHER
DOING SUCH A THING THESE DAYS?
(YES IF HE HAD A GOOD AUTOMOBILE
AND A GOOD SALARY —)

Precios memories. How they linger.

Memories of Reverend Gatewood.

Rob's grandfather, "going back to Alabama."

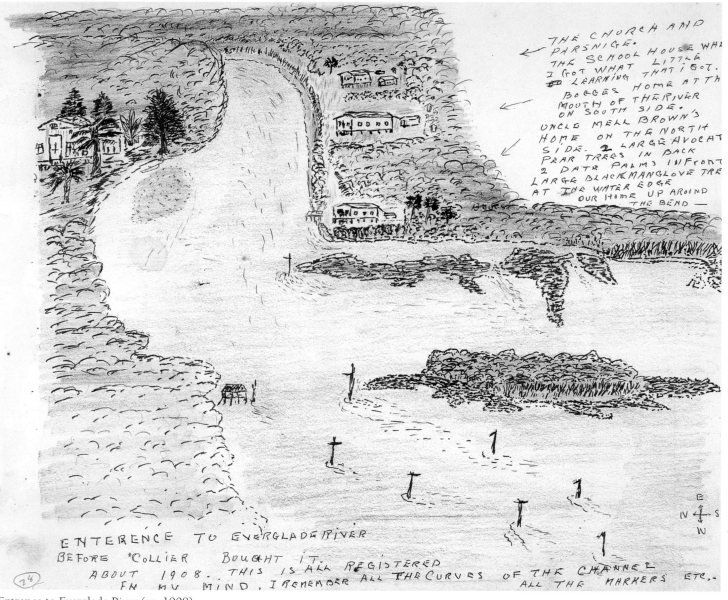

THE CHURCH AND
PARSNIGE.
THE SCHOOL HOUSE WHE
I GOT WHAT LITTLE
LEARNING THAT I GOT.
BORGES HOME AT TH
MOUTH OF THE RIVER
ON SOUTH SIDE.
UNCLE MELL BROWN'S
HOME ON THE NORTH
SIDE. 2 LARGE AVOCAT
PRAR TREES IN BACK
2 DATE PALMS IN FRONT
LARGE BLACK MANGLOVE TRE
AT THE WATER EDGE
OUR HOME UP AROUND
THE BEND —

E
N S
W

ENTERENCE TO EVERGLADE RIVER
BEFORE COLLIER BOUGHT IT.
ABOUT 1908. THIS IS ALL REGISTERED
FN MV MIND. I REMEMBER ALL THE CURVES OF THE CHANNEL
ALL THE MARKERS ETC.

24

Entrance to Everglade River (ca. 1908).

Coming up West Pass,
when porpoises jumped
over the rope.

WEST PASS.

WE ARE ON A SAND BAR
WITH ROPE ACROSS THE CHANNEL
TRYING TO WINLESS OFF THE BAR
THESE PORPOSES COMING DOWN
WOULD NOT GO UNDER THIS ROPE,
THEY ALL JUMPED OVER IT.

I HAVE LOTS OF GOOD STORIES
LIKE THIS. ABOUT PORPOSES.

Little Pavilion Key campsite.

TWO ⬛ DAILY LIFE AND COMMERCE

My father's pride schooner, the little *Bertie Lee,* which he sailed for many years when I was a little boy, was named after the first girl in the Storter family, Uncle George's firstborn, born in Everglade about 1890. I remember this boat well and went to Key West many times on it.

The *Falcon,* a Key West sloop, had been in a fire and had sunk. My dad rebuilt it, cut it in half, and made a schooner out of it, and in 1898 he put in the first two-cylinder Globe gasoline engine that Everglade had seen and renamed the *Falcon* the *Bertie Lee.*[1] He gave every person in Everglade a ride on the schooner the first Sunday after he rebuilt it. The boat depended on the wind, weather, and tides. Lots of times we had no wind, and at other times too much. In some tight places we had to use poles to help.

Although my dad farmed with his father and brother, he was mostly responsible for operating a passenger and freight service along Florida's west coast. Because Everglade was so isolated—there were no roads at this time, only rough trails—supplies had to be shipped in from Key West and Tampa. The

Bertie Lee, a fifty-foot schooner, usually left Tuesday and returned on Friday. Trips to Key West found her loaded down with salt mullet, hides, papayas, sweet potatoes, guavas, eggs, cabbages, oysters in season, chickens, pumpkins, syrup, avocados, limes, and even hogs. In 1896, before the schooner had a gasoline engine and two years after I was born, my dad received a contract for $110 a year to carry the mail from Everglade to Key West. Prior to that he carried the mail from Chokoloskee to Key West for $3.50 a month. When he would leave Everglade and go to Chokoloskee to pick up the mail to take on to Key West, no matter what time the clock said, Mr. McKinney [the Chokoloskee postmaster in 1891] refused to release the mailbag until the rooster crowed.

I accompanied my dad on many trips. Sometimes he would let me steer the boat when the weather was good. I'll never forget the mixture of smells that came from the deck of that boat when it was loaded and ready to sail. Families all over the Ten Thousand Islands brought their goods to my dad to be taken to the Keys and sold. They came on Monday, and by Tuesday the schooner would be loaded and ready to sail. Back then there

was a family on every island that could be lived on, and everyone had something to sell. Aunt Toggie Brown, Mama's sister, would send her hot pepper sauce and sell it for fifteen cents a bottle. They were the same peppers that were stuck to our tongues when we said a bad word. We'd pick guavas and get fifty to seventy-five cents a box for them. Joe Wiggins would send cabbages he grew. A man from Onion Key brought figs and grapes to sell. Mr. McKinney sent redbirds; the Key West market paid fifty cents to $1.25 each for the birds. We carried five-hundred-pound boxes of salt pork, which sold for ten cents a pound. A lot of this salt pork went out to Deep Lake Grove for the fruitpickers there.

Late at night down inside the cabin when I should have been asleep I'd hear papa singing "Amazing Grace" or another hymn and it seemed like he sang all night long. I'd hear the booms squeak, the noise of the water splashing against the sides of the boat—it was beautiful music ringing in my ears. Sometimes we would have a long heavy trolling line with a big babbitt squid for bait, and often we'd catch large kingfish. We had it tied to the back of the

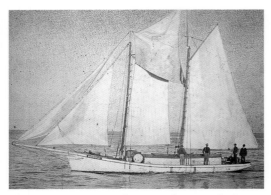

The *Bertie Lee,* prize schooner, photographed in Key West by Mr. Hunt.

boat with a big bowknot in it. When a fish hit it the knot would pull out, and almost every time there was a kingfish on it.

I remember Captain Horr riding on the boat once from Marco to Fort Myers. He had been a United States marshall. He came to Horr's Island mostly for vacations with his wife, son, and daughter. Someone managed his orange grove and pineapple plantation — he had about forty or fifty acres of pineapples. At one time there were quite a few houses there — Howard Helveston's family lived there once, and Howard got in a fight and cut the throat of a black man. Jack Helveston rented the island for a time and buried the black man in the pineapple patch there. Captain Horr was trolling on my dad's schooner and caught a black grouper and made chowder called Horr's chowder. We hauled two loads of pineapples for him. His son-in-law ran the farm and had a boy the same age as me. He wanted me to spend the night. They had two chihuahuas, the first I had ever seen. Captain Horr once tried to raise goats on Panther Key, but the panthers ate them all.

Sometimes my dad would make a trip to Tampa with a load of gator hides and bring back supplies for Uncle George's store — groceries, hardware, dry goods, gasoline, kerosene oil.

The *Bertie Lee* caught fire on one trip to Key West when a careless worker spilled gasoline on her deck. The cabin was completely destroyed and had to be rebuilt. Once we towed a sick whale into Chokoloskee from Cape Romano. The steamer *Olivet,* on its way to the Keys from Tampa, hit the whale and damaged a propeller. The *Bertie Lee* towed the whale to Chokoloskee, where the oil was extracted. A man got down inside the head of the whale and brought it up by the bucketfuls. Another time another captain ran the boat on the Harbor Keys rocks and wrecked it. They filled it with empty whiskey barrels to keep it afloat and brought it back home and rebuilt it.

There were lots of exciting things like squalls and bad weather. I remember dodging a water spout. Dad had to change his course to miss it.

On one trip Dad sent me to the rear of the boat for some sugar. I stubbed my toe and fell, grabbing hold of the railing just in time. I hung for minutes with my feet dangling over the water. Finally my brother missed me and yelled "overboard" just in time. Dad brought me safely back on board. Our mama often said it was miracle that not one of us had been lost or drowned. But cousin Bertie Lee's son Joseph did drown on his second birthday. His grandma was watching him and went to the post office. He slipped out. Someone remembered seeing him near the dock. My dad used a grappling hook to search the river. When they pulled his body from off the bottom of the river the postcard he had gotten from his mama at the post office was still clinched in his hand.

KEY WEST
BEFOR RAIL ROADE—
I REMEMBER—IT—
NEVER AGAIN SUCH A
SCENE.
I REMEMBER THIS. WHEN I WAS A BOY MANY YEARS AGO.
ALL FISH ARE KEPT ALIVE. IN WELLS IN THESE BOATS. THEY ARE CALLED (SMAREYS)
AT KEY WEST - MANY YEARS AGO.
BUISNESS WILL START ABOT 3-30
THIS DOCK WILL BE SO CRODED YOU CAN HARDLY WALK
THIS IS THE FISH DOCK — WHERE FISH WAS SOLD FRESH. DIPPED UP AND CLEANED ALIVE
THIS WAS A VERY BUISY PLACE ABOUT 4 OCL, PM.
HUNDREDS OF CUSTOMERS - EVERY AFTER NOON.

Smackeys at the fish dock in Key West.

The smells of the Key West dock reached the deck of the *Bertie Lee* long before we docked. I remember Papa yelling around ten o'clock the next morning after sailing all night, "All right, boys, see who can see Key West first!" During my childhood every family had some kind of sailboat. Lots of families would cut buttonwood and carry it to Key West, as everybody used wood-burning stoves for cooking. A lot of salted mullet taken to Key West was shipped to Cuba, which was a good market. Everything was done by boat, sailboats and steamships, and the Key West harbor was full of hundreds of boats.

We would sail all night, and the next day we would be looking for the smoke of the cigar factories. We'd sail for several hours and then see little "smackeys" fishing for grouper, grunts, and snappers for the fresh fish market.[2] We would then pass where the navy battleships and Coast Guard boats docked. Mallory Steamship Dock was next, where we would see big steamers most every time. It was exciting seeing the steamboats headed for New York, as these did not come our way in the Ten Thousand Islands. We would have to look straight up twenty-five or thirty feet to the top of the hull. Then we'd pass Sweeney's or the banana dock, where the schooners would be unloading bananas, and we would always get a dingy load of bananas that were too ripe to ship to other places. We ate so many we didn't care to see a banana for a while. Sweeney's Dock had a restaurant where you could buy a hot meal—mostly soup and black coffee—for fifteen cents. It was also called Hobo Place because so many drunks hung around and slept right on the dock. I walked on the dock at night and could hardly walk without stumbling over a sleeping drunk man. Sometimes these hobos would plunder the schooners, searching for food and money. Once a hobo got on board the *Bertie Lee* and was chased off. Papa always slept on the boat because of this.

The sponge docks were next. Sponge fishermen would sometimes pay us fifty cents to fill corn sacks with sand until our dad discovered the sand was being put in the sponges to make them weigh more. Curry's Dock had a large general store where you could get lumber, hardware, dry goods, and groceries. My papa would buy Nabisco cookies in tin boxes—the cookies were displayed behind glass. This was where I spent any money I had. Grandpa bought hardtack and always had some in the galley. There were also huge turtle pens underneath Curry's Dock and a canning plant for turtle soup. I'll never forget the smell of turtle soup cooking. There were hundreds of turtles—loggerheads and green turtles. I saw one small schooner come in with turtles all over the deck and lots of them down on the inside. The turtles, on their backs, would have holes cut in their feet, and they would be tied together to keep them still. At the fish dock there were fish smackeys and sailboats with wells in them with live fish of all kinds. There were lots of these boats, and we would see them coming and going around four o'clock. There was hardly elbow room on the dock, as people were buying fish for supper. The fishermen would reach down inside a well and come up with any kind of fish you wanted, hit it on the head, and then clean it for you. There was no string for them to use, so they used silverleaf palm leaves for tying through the gills of the fish. We unloaded and loaded at Sweeney's Dock, which was about three hundred feet long and had a roof over it. The *Magnolia*, a big schooner, was always at the end of the dock, loaded with dry goods.

My papa would hire help when we got to the dock to help unload. He had his favorites, and when they were available he used them. Before we'd ever dock I'd hear, "Hire me, Captain, hire me!" I saw one Negro grab the stern line and one the bow line and get into a regular fistfight over who got there first. Papa would say, "I can't hire you both," and he usually had to settle the argument.

The next morning at ten o'clock there was an auction. We would take our stuff there on a big flat wagon, and my papa kept a record. One man we knew was using one hundred pounds of salt and one hundred pounds of sand broken up to put in the barrels of fish, and got caught. Later, after he got converted, he went to my uncle George and confessed he had changed his name in order to sell his fish. Uncle George told him to forget the past.

On one trip to Key West there came up some real bad weather. It rained and blew almost hurricane force all day. We had to head straight into the storm and go through the Keys from Long Key to Key West. Papa was sick with a really bad cold, but luckily we had a mate along. His name was John Curry, a one-armed man who was raised in Key West and knew those inside channels. He told Papa to just turn it over to him and he would take us on to Key West, and he did. Sometimes we would drag bottom, but we never got stuck. Old Man Curry was a gator hunter and would scull his boat with one arm and one oar for miles.

"AUNT EM FED HER KIDS . . . ON SWEET POTATO VINES." ROB

In 1908, my uncle operated a sawmill using the stationary motor from the cane mill to run the saw. They cut cypress lumber until the supply became exhausted.

One summer over ten thousand alligator hides were shipped out to Tampa from my uncle's general store. He took in hides and furs from the Seminole Indians. When I was a boy I spent lots of time around at the store watching the Indians trade and measure their gator hides. Sometimes this would last all day. The Indians would sometimes come in several families. I saw as many as fifty in a bunch, and they would spend several days trading. They would stay in a big shed where my uncle built boats. Sometimes my dad would give them rides on the schooner. At times they would buy a cow from my uncle and have a real feast for several days. Sometimes they would order things out of the Sears Roebuck catalog. They made sofkee and cooked it in a big iron pot.[3]

Cutting Cypress by hand for Uncle George's sawmill.

They would gather around and sit flat on the ground with their feet folded under them. The women did the cooking, but only the men ate; I never saw the women eat.[4] They had a homemade dipper that was used as a spoon. They would joke during the meal; we couldn't understand their language. They would get drunk—most of them at one time, except for about two of them who would stay sober to keep order. They never got out of order. They just seemed to be happy and having a good time singing. I don't know where they got their whiskey. When the celebration was all over they would leave together at one time with their canoes loaded with groceries and supplies. Charley Dixie, a half-Indian-Negro, who was not liked by the Indians and later was murdered by one, told me one day that he had killed sixty deer in one day to make buckskins to sell to my uncle who paid cash for the skins.

Many schooners tied up at the dock. There was a man from Punta Gorda named Wotitzky.[5] Every year in the late summer or early fall he would load his sloop with dry goods and go from place to place to sell his goods. I can remember him as he would park or tie up at our dock for several days. He always had lots of children's clothes, shoes, and hats. Mama would buy things for us kids.

Charles Gardner and his wife—we called her Aunt Em—sold gator hides to my uncle. Charles often would not make it home with

Seminoles living in Everglade.

the money. One time we heard she slapped him in the face when he went off and left the family without anything to eat. Aunt Em fed her kids on sweet potato vines. As Bill, her boy, got older, he and Aunt Em would trap coons at Half-Way Creek to sell to Sears. She was an excellent trapper and could stretch hides better than anyone I ever saw. They would look almost square. She also hunted alligators. She was a good shot, but she would row the boat and Bill would shoot the alligators.[6]

My uncle George wanted to send Minnie to school, but Aunt Em said she was too weak because she had hookworms.[7] Minnie's brother Bill offered to teach her to read and write, but only out of the Bible. She refused.

When Bill had been converted at Chokoloskee he had just learned how to make moonshine. In later years he convinced the church we went to in Naples that we were supposed to use real wine for communion. He made

the wine for the service, and my son, Lem, and his friend Lorenzo Walker got into it, and that ended that.

I remember one real rainy season when the water was high and the Everglades was full. My uncle George and a party made a trip across the Everglades. I don't know how many of them there were, but there was quite a crew of them. I remember the head man's name was Dimock; I suppose he was trying to get some data for a book.[8] They had an Indian for a guide and made a trip across the Everglades to Miami from the head of Shark River.

I remember another old fellow, a Norwegian named Nick Johnson. He would often tie up at our dock too. He was a peddler of anything he could buy and resell. He made trips down to the Keys trading with the people that were making the overseas highway to Key West. He was the best with an accordion. We enjoyed hearing him play. We bought a black pineywood rooter pig from him just before the 1910 hurricane. We named him Nick. Nick was a good pet. When the hurricane came it was one of the worst. It carried Nick off. And we didn't see him for several months. We thought we would never see him anymore, that he had drowned, but one day Nick found his way back home, and how glad we kids were to see him. He was very poor—he had almost starved. But we gave him the best and got him in good shape again.

My aunt Penny Brown lived at Punta Rassa and we visited her many times when my dad took alligator hides to Tampa and brought back supplies for Uncle George's general store. My dad would put Mama and us kids off at Punta Rassa and pick us up on his way back. When I was ten years old I remember seeing cowboys driving the cattle into the pens and seeing them loaded onto the three-masted schooner *Doctor Lykes* to go to Cuba. One old cowboy said the Cubans were crazy about bull meat. It was not unusual to see several thousand bulls and steers. They would keep them separate because the bulls would fight the steers. Captain Nick Armeda ran the schooner. I knew him very well for many years. The schooner belonged to Lykes Brothers. My aunt would board the cowboys when they came with a herd of cattle. They ate grits, lima beans, and cornbread. Sometimes they would kill beef and salt and dry it. It was hard to eat and salty. Aunt Penny would soak it in water first so we could eat it. It was sure a sight to us boys when we heard the whips popping and the cowboys hollering. I and my cousin Frank Brown would climb up a black mangrove tree and watch the herd go by.

The cows were herded into pens and then driven down a narrow slip and forced on the boat. Some balked, but the cowboys would twist their tails and punch them with a sharp stick. The chute that led to the boat was made of pine boards that were sometimes rotten, and a cow would [sometimes] fall through or his leg would get broken and would have to be killed. If someone was there they could have the cow—it was good beef.

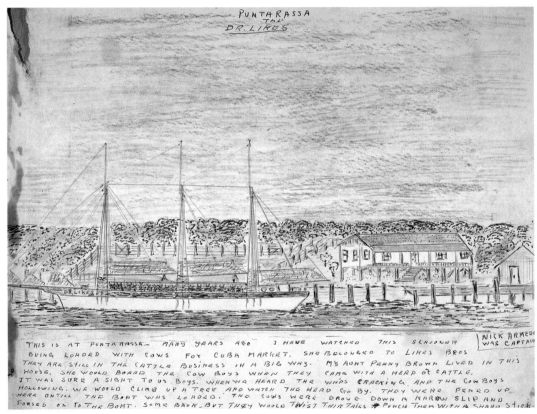

The *Doctor Lykes* being loaded in Punta Rassa with cows for the Cuban market.

Pat Roll and Dode Russell helped take cane from Half-Way Creek to Everglade for my dad.[9] I helped in the cane and syrup-making, too; I could feed the mill to make the juice and also boil the syrup. At the syrup-making season, you could smell the syrup cooking as soon as you got on the river. Most of the cane was hauled from Half-Way Creek on a big barge or lighter before motor days. This forty-foot scow was moved with poles and loaded with green cane. I remember a little story of a happening at Everglade. My dad had just returned from a trip to Tampa with a load of groceries and dry goods for my uncle George's store. They had a steam boiler on the stern of the boat; it was a little on one side. When they unloaded the freight [all but the boiler] the boat tilted and the boiler rolled overboard. I remember they had an awful time getting the boiler out of the river. They worked several days and finally got it out and put it where they intended it to go. I think it was for the cane mill.

Pat and Dode were hired hands for the Storters. Pat, a Negro whose name was Erskine Roll, had been a stowaway on the *Bertie Lee,* and once in Everglade the five-year-old begged to stay with my uncle. He had run off from his parents in Key West, so George went to Key West and asked if the boy could live with them. Pat became an expert syrup-maker. George bought him a banjo, and he was a very gifted player. He was a good boy, didn't smoke or drink, but he was not allowed to go to school or eat with the family. There was a table in the kitchen for him. He eventually went to Lostman's River, married a white woman by the name of Annie "Shoog" Hamilton, then he moved to Key West. Pat and Annie did laundry for families in Everglade for a time.

Dode was killed when a boat exploded while hauling gas to Everglades right after Barron Collier bought the town. His brother, Pomp [Ronie Russell], pulled a cord to ring a bell and a spark came and caused the explosion.

The *Ethel Q* from Fort Myers was the last schooner that my dad owned. He traded a nice thirty-foot family launch as a down payment on the *Ethel Q.* Papa cut her down the middle and lengthened her and made a sure enough boat out of her.[10] The *Ethel Q* did not have a motor, but my dad had one installed and rebuilt the cabin so it would be more suitable for hauling supplies. He no longer carried the mail to Key West when he owned the *Ethel Q* because the railroad had been built. But he still made trips bringing supplies to and from my uncle's store. The *Ethel Q* was the first schooner to carry iced fish out of Everglade. The *Ethel Q* made trips to Chatham River to the old Watson cane farm for a load of cane to be delivered to Moore Haven for some of the first seed cane for Okeechobee. I, along with my dad, my cousin Bill Stephens, and my brother Wes, left one morning for Chatham Bend, and on the way down we sailed until we got in the river. Then we got on lots of mud bars and had an awful time getting up to the cane farm. We had in tow a small scow. Bill was always fishing with a rod, trolling for something, and often he would catch a fish of some kind and it always came in handy at mealtime. But we finally made it up to the cane farm and started loading the cane as the man and his helper would cut it. We toted it out on our shoulders to the boat, stacking it down in the cabin until we loaded the schooner.

We were several days on this trip. We ate up all our food. Maybe we had an onion and a potato or two [left]. We left the cane farm and had a worse time getting out of the Chatham River, but we got out just before sundown with everybody almost starving. Dad said, "Boys, it's a long way home, but under us are lots of clams." The tide was pretty high by this time—about shoulder deep. He told us if we would dive for a few clams he would fix us a chowder. Overboard we went. We could feel clams most anywhere you could put your foot, but getting one in that deep water was a problem. One of us would find a clam with his heel or toe and one would try to dive and dig it out, but that was sure a job to get one. The current was strong and the water deep. But anyway, we managed to get about six big bull-nose clams. We opened them and Papa made the best clam chowder I ever tasted. A few days later we carried the cane up the Caloosahatchee River to Moore Haven, and that started the cane business in that part of the country.

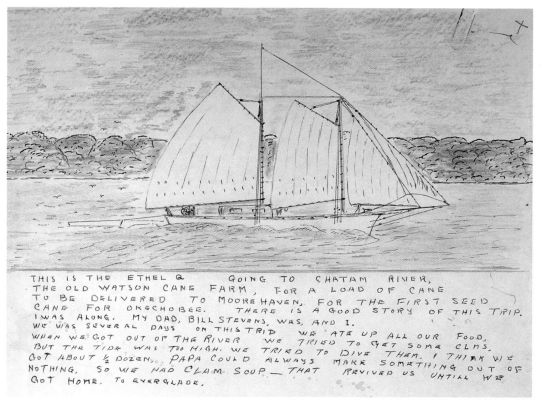

THIS IS THE ETHEL Q GOING TO CHATAM RIVER, THE OLD WATSON CANE FARM, FOR A LOAD OF CANE TO BE DELIVERED TO MOORE HAVEN, FOR THE FIRST SEED CANE FOR OKECHOBEE. THERE IS A GOOD STORY OF THIS TRIP. I WAS ALONG. MY DAD, BILL STEVENS, WES, AND I. WE WAS SEVERAL DAYS ON THIS TRIP WE ATE UP ALL OUR FOOD, WHEN WE GOT OUT OF THE RIVER WE TRIED TO GET SOME CLMS. BUT THE TIDE WAS TOO HIGH. WE TRIED TO DIVE THEM, I THINK WE GOT ABOUT ½ DOZEN, PAPA COULD ALWAYS MAKE SOMETHING OUT OF NOTHING, SO WE HAD CLAM SOUP— THAT REVIVED US UNTILL WE GOT HOME. TO EVERGLADE.

The *Ethel Q* on her way to the old Watson cane farm.

LUNCH TIME   — COUSIN BILL & BRUTIS —

Fishing and hunting sustained us. Deep Lake Grove was a favorite hunting area. There were plenty of deer. Hundreds of acres of grapefruit went to waste [in that area] because there was no way to get the fruit out. When Walter Langford of Fort Myers bought Deep Lake Hammock he put in a dummy railroad from Deep Lake to Everglade. I recall about fifty Negroes being brought in to pick the fruit. My brother Claude was the railroad engineer. My grandfather and uncle drove the first spike in laying the rails. The men would labor all week for nine dollars and then gamble on the weekend with the babbitt [metal] money they had been paid. The men ate mostly lima beans and sow bellies [white salted bacon]. On one trip into Deep Lake I heard so many baby birds crying that I lost my desire to hunt.

Birds were plentiful, especially the plumes. At one time they sold for five dollars apiece and became so valuable that hunters quit hunting other animals and went after the plumes. Finally, the state passed a law forbidding the killing of these birds. When the law was passed in 1901 it was not immediately successful, but as fashion changed and the law broadened to forbid the importation of the feathers the killing stopped. My dad told us of a man that lived in Everglade that took a trip to Honduras and found the plumes so plentiful that he wrote his sons to join him and bring plenty of ammunition. The letter was intercepted and read by two men in Key West. The men went to Honduras and killed many of the birds and made a mattress out of the feathers and tried to come into the country with one of the men pretending sick on the mattress. They were caught.

When I was a boy there were no game laws, but when Collier bought Deep Lake Grove the turkeys were protected and no shooting was allowed in the grove.[11] The wild turkeys were very tame and there were several hundred in the grove. One evening on the porch of the big house where the white foreman stayed I counted eighty-six turkeys as they would fly up to roost in a cypress head by the kitchen.[12] Some of the turkeys would eat corn with the chickens.

Deep Lake at sundown; grapefruit trees.

ABOUT 1912

THIS IS THE END OF THE
DEEP LAKE ROAD. 12 MILES TO DEEP
LAKE. THIS IS THE HEAD OF THE RIVER.
4 MILES DOWN TO EVERGLADE FOR SUPLIES
BACON, BEANS, & FLOUR. THIS OX TEAM COMES IN
ONCE A WEEK, EVERY TOESDAY. THE SUPLIES ARE
LOADED ON THIS OX TEAM AND HAULED TO DEEP
LAKE. I HAVE WALKED THIS MANY TIMES WHEN
I WAS A BOY. WAS GOOD HUNTING, PLENTY DEER.

Oxen team heading to Everglade
to haul supplies.

About twenty-five clam diggers camped while digging clams on Little Pavilion Key at every low tide. This was done for several years and furnished all the clams for the Burnham factory at Caxambas, then for Doxsee Canning at Marco.

W. D. Collier invented a dredge that put the hand diggers out of business.[13] The clam canning was finally moved to Naples until the clams played out and the clam business ended, and it has not come back. The clams died for some unknown reason, and they tried to use conchs, but that didn't pay. There are many theories, and no one knows for sure, but my guess is red tide and hurricanes, and the dredge covered up a lot of them, and a lot were canned up. This clam bed produced for over fifty years for the canning factories. Well, this was an industrious time. A good digger could dig about twenty-five bushels on one tide, for twenty-five cents a bushel.

The *Clara*, a clam boat, took over five hundred bushels of clams to be canned at the Naples factory. One morning as the sun was coming up and I was coming out of Indian Key Channel, Claude and I were on our way to Shark River with a fishing party in a yacht. I told Claude it looks like there's a man stand-ing on the water out there, and he told me to look through the binoculars. When I looked I told him it was the clam boat and it had sunk and a man was standing up and another man was sitting down on top of the pilot house. So I turned loose in a rowboat and ran out there and rescued them and carried them in to Ever-glade. They were two happy men. They had been on top of that pilot house all night, and it was a cold night. They said at high tide there was only one inch of dry space at one corner of the water. She sank later that morning at ten o'clock.

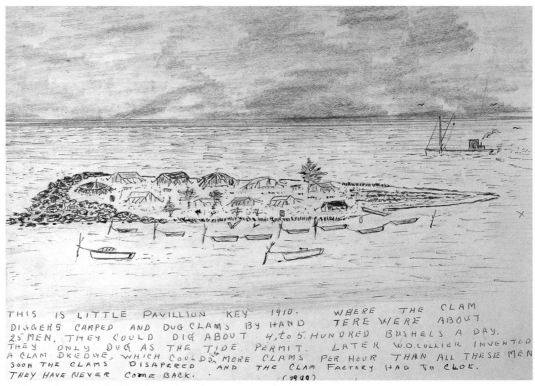

THIS IS LITTLE PAVILLION KEY 1910. WHERE THE CLAM DIGGERS CAMPED AND DUG CLAMS BY HAND. TERE WERE ABOUT 25 MEN, THEY COULD DIG ABOUT 4, to 5. HUNDRED BUSHELS A DAY. THEY ONLY DUG AS THE TIDE PERMIT. LATER W.O.COLLIER INVENTED A CLAM DREDGE, WHICH COULD MORE CLAMS PER HOUR THAN ALL THESE MEN SOON THE CLAMS DISAPERED AND THE CLAM FACTORY HAD TO CLOE. THEY HAVE NEVER COME BACK. (1940)

Little Pavilion Key, 1910, where clam diggers camped.

Clam digging dredge.

THIS CLAM DIGGING DREDGE COULD DIG MORE CLAMS
PER HOUR THAN 25 MEN COULD BY HAND BUT IT BROKE UP
LOTS OF CLAMS. NOW THE CLMS ARE GONE ONLY DEAD
SHELLS WHERE THERE WERE MILLIONS OF GOOD CLAMS
THE LAST CLAMS WERE CANNED AT NAPLES, BUT IT HAD TO CLOSE
AND GO OUT OF BUSINESS AND THE OLD FACTORY IS A STORAGE
HOUSE FOR BOATS.

                                              W.D.
THIS DREDGE WAS PATENED AND MADE BY CAPT., BILL^COLLIER. MARCO
                                                          FLA
THE CANNING FACTORY - FIRST AT CHXAMBAS, THEN TO MARCO,
AND LAST NAPLES, BUT THE CLAMS — DISEPERED.
          MANY IDEAS WHAT HAPPENED TO THE CLAMS —

(52)

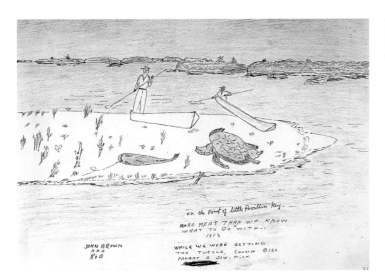

Fishing off Pavilion Key Point in
1913, where a turtle and jewfish
were caught.

on the Point of little Pavilion Key.

MORE MEAT THAN WE KNOW
WHAT TO DO WITH—
1913

JOHN BROWN
AND
ROB

WHILE WE WERE GETTING
THE TURTLE, COUSIN BILL
CAUGHT A JEW-FISH.

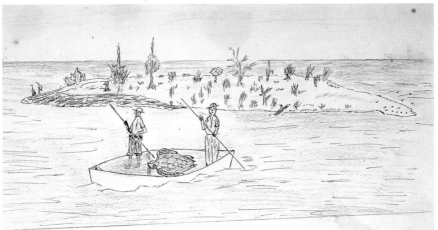

Fishing off of Little Pavilion Key.

ON THIS LITTLE ISLAND IN 1909 ABOUT 25 MEN CAMPED
THEY DUG CLAMS BY HAND, ABOUT 500 BUSHELS A DAY @ 25¢ per bush
THEY WERE HAULED TO CHXAMBAS AND CANNED FOR N.Y. MARKE
LATER WERE DUG BY DREDGE. AND CANED AT NAPLES
BUT NOW CLAMS ARE GONE

(ROB & COUSIN JOHN)

LITTLE PAVILION KEY
JOHN & ROB   1913

"WE KEPT OUR DOORS OPEN DAY AND NIGHT." MARILEA

The ice plant opened in 1922, and Sidney Griffin had the first ice delivery and served the town quickly, making deliveries himself. He put ice in the boxes at the houses whether or not the people were home. Sometimes dead bodies were kept in the ice plant until they could be moved to Fort Myers. When delivery stopped, every other day we would have to go get ice. One day my girl Olivia walked in on a woman being kept there awaiting burial. This woman, Florence, had been married to a man that was many years older than her and he was very mean to her. She was only thirteen when she married, and she got divorced and remarried this older man about four months later. We knew her parents, Charley and Lillie Johnson, and visited them on Sunday afternoons.

The women often sat with each others' kids. There was no doctor or immunizations and when eight of the children came down with whooping cough during Christmas, mothers would take turns watching them at night so the others could go to a revival at the schoolhouse. They spread newspapers all over the schoolhouse floor in case the children there "coughed up."

We kept our doors open day and night. Nobody thought to lock up. There was no vandalism. Young people picked fruit, flowers, and coconuts from the winter residents' houses in the summer, but didn't think to damage the property.[14]

In 1929 we got electricity, as the first electric plant opened, but not many had this luxury.[15]

"AN OLD SPANISH MAP . . . SHOWED AN ISLAND WHERE THIS CANAL ENTERED THE GULF." ROB

Mr. Ed Crayton of the Naples Improvement Company gave me permission to build a house at the mouth of the old unrecorded canal on the bayside.[16] The canal had at one time reached from Naples Bay to the Gulf of Mexico. I talked with Speed Menefee, first mayor of Naples, about this canal, and Speed told me that he saw an old Spanish map that showed an island where this canal entered the gulf, and he supposed the Indians used it, as it was the only outlet to the gulf at that time. After I moved to Naples I helped fill in the canal. Homemade Model T dump trucks moved sand from the beach to the canal site. We were paid ten dollars for eight hours of work. Forrest Walker [next-door neighbor], Cass Pollock [Marilea's stepbrother], and my brother-in-law Charlie Summerall helped. I recall Charlie filled his truck with long slow sweeping shovelfuls, and I filled mine with quick short ones. Charlie had his truck filled up long before mine every time, no matter how hard or fast I worked.

STORY ON THE BACK

Rob Storter 83

Hauling sand from the beach in homemade model T trucks for the construction of Naples.

Forrest Walker's brother Rosco helped dig the Tamiami Trail at the head of the Fakahatchee River. I carried supplies to him from Naples by boat up the Fakahatchee, which was the end of the trail at that time. My uncle George helped blaze the right-of-way for this road. Workers were paid twenty cents an hour.

Many families came to the area to work on the trail, and many stayed in an area called Tent Village, and later Poverty Row, including Marilea's brother Charlie and his wife Zola. Tents about sixteen feet by sixteen feet housed a family, and a nine-foot-by-ten-foot tent was for a single man. The Jake Jones family, from Georgia, lived in the village located near the Catholic church. A son, Stephen, fell in love with Bea Bickford, as she was a tent neighbor, and he would trim the girl's dark red curls as she sat on a nail keg. Her mother ran a mess hall for the workers in the village. She would drive a Model T to Fort Myers to buy huge supplies of food to feed as many as forty people. Sometimes they would buy venison and wild turkeys from the Indians. Her husband was a bridge builder.

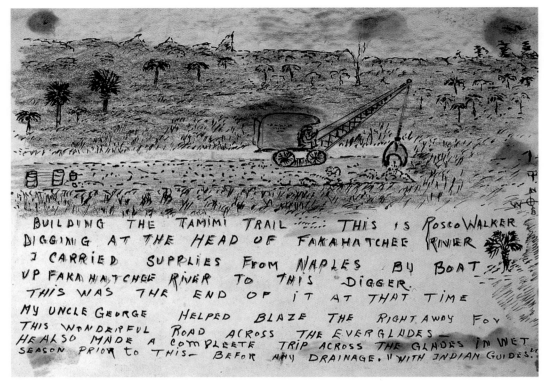

Building the Tamiami Trail.

In 1935 we moved to our second home in Naples, on Goodlette [Pulling] Road. I asked Mr. Crayton if I could have the lumber from an old fifty-foot houseboat at Naples Bay. Naples's first mayor had lived in it, then other families. That's the wood I used to build the Goodlette Road house. When I was tearing it down a two-by-four hit me, caught my belt and threw me up in the air and down on my back, knocking the wind out of me. I got up and finished tearing the boat down after Ernie Carroll rushed over and asked if I was all right.

When the depression ended the tourist and guiding business boomed. The building business boomed. Wilbur and I helped build the first tourist court, owned by Dan House. We were working on the main office, a two-story building, when Wilbur was struck by lightning. It was about one-thirty in the afternoon and we were completing the top of the roof with metal shingles, and I caught Wilbur in my arms and began praying. The bolt had hit him on the head and torn his cap, ruined his teeth, and almost welded the tin snips to the shingles. It seemed like hours before they could get him down, but it was only minutes. I remember holding him in my arms and putting my hand over his heart. I felt nothing. Wilbur's heart did not start beating again until we got him down on the ground. For days we did not know whether or not he would live, but he slowly recovered after many weeks.

One day at the Bay Dock I hooked the boat onto a rope and pulled it out with the car. The ring pulled out and flew and hit Ed Townsend in the jaw and cut and broke his jaw in two. Immediately, blood poured, and he began to pray and ran to his father's houseboat. His wife and brother sat up all night with him. I had to go outside once because the smell of blood made me sick. There was no doctor around. He got okay after that.

Lightning strikes brother Wilbur.

Ted Smallwood's trading post and store.

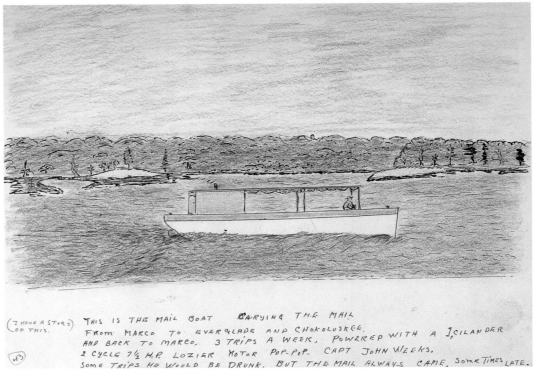

The mail boat, manned by John Weeks.

Loggerhead laying eggs.

THIS WAS BACK IN THE DAYS
WHEN SISTER WINNIE WOULD
CROSS THE RIVER IN A WASH TUB.
AND CLIMB ALL THESE GUAVA TREES.
— THE HOUSE ON THE RIVER BANK
WAS CALLED THE SUGAR HOUSE
THERE WAS A BARREL OF SYRUP THAT
TURNED TO BROWN SUGAR + ROCK CANDY
THE ROACHES WOULD EAT THE SUGAR AT
NIGHT. AND WE KIDS WOULD TAKE OVER
NEXT DAY.

PRECIOUS MEMORIES
WHERE MAMA + PAPA
RAISED 9 CHILDREN

THE STORK WAS WELCOME

WOOD, STOVE FOR COOKING
KEROSINE LAMPS. FOR LIGHT
WASH RUB BOARD — FOR WASH DAY
RAIN WATER.
NO RADIOS — OR T.V. NO CARS —
BUT OH WHAT FUN WE HAD
A 24 QT. ICE CREAM FREEZER WHEN
WE COULD GET ICE
OH WHAT A TIME. NO WORRIES
PLENTY TO EAT —
TIME TO VISIT EVE

— EVERGLADE — BY MEMORY — OUR OLD HOME AS IT WAS — (1906) —   WHERE I WAS BORN 1894

Everglade, the old homeplace.

THREE ✦ FAMILY AND FRIENDS

"I KNEW WHEN GRANDMA DEES CAME THAT MEANT A BABY WAS COMING." ROB

Allen's River was our home. The house was twelve feet wide and twenty-four feet long with two rooms and a kitchen. My father said before my mama turned fifteen the house was finished, and just after she turned fifteen he took her to Key West and married her. Two years later he bought property from his brother, moved the house there, and rebuilt it. Five years later he added several twelve-by-eighteen-foot rooms onto the place. They raised nine children there.

My dad fell in love with my mama when she came to stay with Uncle George and Aunt Nannie. Her name was Nancy Stephens and she had been born in Apalachicola, Florida, in 1876. One day as the girl and her parents rowed to Fort Myers for their weekly supplies, a rainstorm drenched them and her mother caught cold and died. After her death Nancy and a brother and two sisters went to live with an aunt in Carrabelle. No one was sure just why their father came and got them, because he was such a poor provider, but he did. Aunt Maggie, Mama's sister, begged and cried, wringing her hands when her father came to get them. When he came to Everglade he would lie on the bed, chew tobacco, and

spit into a can. He did seem relieved that Nannie and George took Nancy to live with them. When my dad told George and Nannie about his intention to marry Nancy, my aunt would not hear of it unless my dad provided a house. "She's had enough hard times, and if you want her you have to build her a house," Nannie said.

Most of us were born right in our house on Allen's River. I recall Aunt Toggie Brown being present at a birth and saying, "It's another red-haired boy." I remember Dr. [Solomon] Green, who delivered me. He was the only doctor in our neighborhood in those days. I was named for my dad and the famous general, Robert E. Lee, and wore dresses until I was three years old.

I was born during the 1894 hurricane while my dad was weathering the storm down at Shark River on his way back from Key West. About every two years or so Dr. Green would come to our house and stay about two weeks and leave another baby. He had a long white beard. He was a good friend to my parents and everybody in that part of the country. After Dr. Green died Grandma Gandees, or Grandma Dees as everyone called her, was the midwife when a baby was born. She had

a limp from a broken leg that grew back crooked and no one forgot the story of the local boy who made fun of her and not long after got a limp just like hers. She would move in and stay two or three weeks with a family. I knew when Grandma Dees came that meant a baby was coming. She had a son named Ocean Gandees. At her house the floors were spotless. When she wasn't there anymore Mr. C. G. McKinney of Chokoloskee delivered babies. They said he was good in this business. He also pulled teeth. He had a little store at Chokoloskee—once the post office.

If we had an aching tooth, Mama always had a remedy until we could see Mr. McKinney and without any painkiller he would yank it out. Once I had to go to Key West to have a tooth pulled and I almost came out of the chair. The only good thing about it was I got to miss school for a week to go down there. In later years, a traveling dentist would come along with a foot-peddling grinder and some filling and he would take care of our teeth. He would go from house to house to see that the whole community was taken care of. I always

35

Nancy and R. B. with their sons Claude, Rob, and George.

Church portrait in front of the Everglade schoolhouse, before the 1910 hurricane destroyed the building. Rob is in the center, dressed in black. Rob's younger brother Wilbur is wearing a plaid skirt.

Rob, holding his dad's pocketknife, and his older brother, Claude, in 1895.

liked to be out fishing the day I thought he would be at our house.

Mama used some home remedies, like white bacon or brown Octagon soap and sugar to bring a boil to a head. One day one of the little girls was handling some of Aunt Toggie's hot peppers and started screaming. Luby Hobbs, the hired black man at McKinney's, ran to the chicken coop and got some chicken manure and rubbed it all over the girl's hands and she stopped crying.

One time my cousin Clarence Brown and I were out in the bay diamondback terrapin [freshwater or tidewater turtle] hunting and we went up a little mud creek and I tried to jump across the creek without falling in. I stuck a sharp stick almost through my foot. It bled so much we got scared. Clarence tore the lining out of his cap, and I kept it on my foot to stop the bleeding. That night it hurt me so bad. Mama soaked it with boiling rags. That kept me from having blood poisoning. Papa carried me to Key West to a doctor. He cleaned it out and said there wasn't anything in it. He gave me some heavy white salve and I kept it full of that until I got well.

My uncle Mel Brown used nutmeg on a necklace string around the neck to keep his kids from having croup when they had a bad cold. Tommy Thornton, an old fisherman, told me if I ever got stuck with a poison fish, stingray, or catfish just put gasoline on it and it will stop the pain and not give you any trouble. It works.

This photo of R. B., taken in Naples Bay by Stephen Briggs of Naples, was used in the 1940s to advertise Evinrude motors.

Nancy with seven of her children. *Left to right*: Claude, Rob, George, Eva, Wilbur, Wesley, and Winnie. Missing were Herbert, who was shrimping, and Mildred, the youngest, who died as a teenager.

We grew up surrounded by guava, banana, mulberry, and coconut trees that yielded plenty of fruit for us. All the trees were named. Most of the shell islands had gumbo limbo trees [*Bursera simaruba*]. You can take a limb of it and plant it as a fence post and it will grow—this was done on Chokoloskee Island as landmarkers. In later years when the island was developed and sold in lots, the only way they could tell who was who and who owned this and that was by the old gumbo limbo fence markers.

We had an old cane-grinding mill pulled by a horse. There was a cow for milk. Near the river there was a sugar house, and we could not resist eating the hard syrup or rock sugar off the sides of the barrels. The coons and roaches ate out of it at night. We had a pet raccoon that would dip his paws down into the barrel to get to the sugar, and one day he fell in trying to reach it with his hind leg. We had no refrigerator or electric lights. We had a wood stove, kerosene lamps, and backyard toilet. Mama used washtubs and a washboard and washpots for hot water, and buttonwood for cooking. We never went hungry for lack of food. What appetites we had!

Some of the young boys smoked banana leaves or anything else that would burn.[1] Tobacco came in big plugs about eighteen inches long. It was called "South Down" and it was cut up with a knife and could also be chewed. It had a really strong smell. "Sensation" was another brand that came in pound tin cans. Someone planted a tobacco plant on the abandoned Layne place on Ferguson River. Once, me and my brother George and the adopted son of Hiram Stephens rowed eight miles to the Layne place and stole a tobacco leaf from there, made cigars from the yellow leaves, and floated down the river, smoking. When we neared the Stephens place it was dark. Suddenly, out of the darkness we heard, "Boys, those cigars sure smell good!" I remember I was so sick and scared to go home that I went under a guava tree and sat.

One of my earliest memories was of an airplane that fell a few hundred yards from our house. The Lee County tax assessor had hired the plane to take him to Marco and Everglade when the area was still part of Lee County. Two men were killed, and my dad took the remains of the wrecked plane to Fort Myers, where the flight had come from. My brother

One of Rob's early memories—an airplane that crashed a few hundred yards from the house.

saw the plane fall, and when he got to it he saw two men strapped in the seats, but before he could get to them, the plane burst into flames.

When I was a boy the rivers were full of snook year round. No one ate snook; they said it tasted like Octagon soap. I don't know, for I never tried snook. But we caught lots of them with hand lines and bacon rinds or white rags and later with bright fishing spoons. We caught big ones and roasted them on a fire for the chickens to eat. We thought it made them lay more eggs.

The snook smelled so good. I was tempted to eat some, soap or no soap. We just didn't know how to filet them. But now they are good, so good. I remember Eliza "Pappy" Turner had a heavy twine snook net and caught them by the thousands of pounds, often thirty to forty thousand. Finally catching snook with a net was outlawed and snook were made a sport fish.

We didn't have refrigerators and had to have everything fresh or salted. We would go bird hunting about once a week and could always get a mess of curlews [white ibis] and ducks, which were sure good to eat. Often the Indians would bring venison, fresh and salted. The way we would tell if they were bringing fresh venison was if we could see palmetto fronds, which were used to cover and keep the meat cool. They sold venison ham for fifty cents. When we grew older we would hunt deer ourselves. Fish were always plentiful— just get a little can of fiddlers [crabs] and go

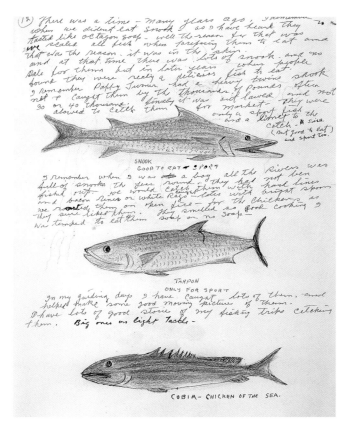

Three types of fish.

down on the river bank. It wouldn't be long before you'd have a mess of sheepshead and mangrove snapper, redfish, or jewfish.

We had a date palm tree near the house. It had sharp, hard thorns on it. I was getting some dates off of it one day and stuck one in my knee just below the kneecap, and it is still there—it feels like a little hard marble. It never gave me any trouble, only at the time I couldn't get it out so I just left it there.

We had to make our own homemade ice cream, and that was only when we could get ice and that was not very often. It was packed in a big sack of sawdust and hauled sometimes for days. Then we would have a party. The whole community of children would get together and play games until the ice cream was ready. We had a twenty-four-quart freezer which belonged to the church, but we called it the community freezer. We made it full, and often there was none left over when we got done. That was lots of fun then, but things have changed now. We generally had only one flavor, vanilla.

I was deer hunting one day and found a fawn. I don't think it was but a day or so old. I took it to the camp. We had a mosquito bar.[2] I turned it upside down, or downside up, as a pen and kept it there for a day or two, fed it with canned milk, and toted it twelve miles and raised it until it was full grown. It would eat anything we ate, raw oysters, as many as you wanted to open for her. She would even eat raw fish and could jump out of a six-foot cistern. She found Aunt Nannie's rose garden and ate all the roses. So she said I had to do something with that deer. Well, I couldn't keep it in a pen. One day, some tourists came along and gave me twenty-five dollars for her. Well, that was a lot of money in those days. I could buy lots of things with that much money, like a new rifle.

"GRANDPA STEPHENS TOOK A MAIL ORDER BRIDE . . . AND THE POOR WOMAN THOUGHT SHE WAS GETTING AN IDEAL HUSBAND." ROB

When I was a boy Grandpa Stephens came to live with us for a while. We all loved him. He was a carpenter and a cabinetmaker. He also would do other things. The first time I saw Grandpa Stephens was when we picked him up in Key West when he came in a Mallory steamer from New York. We went to La Brisa and watched the steamer go by.[3] When he got to Everglade the first thing he wanted was for me to take him to get blue crabs. He had a big pocketknife, which I wanted so bad. I wanted to whittle with it, so he let me, and I dropped it overboard. He liked crabs and oysters and fish. In oyster season he would pickle oysters, boil them in vinegar with cloves and spices. I don't know just how he did it, but when they were put up and ready to eat it was the best thing I ever tasted. He never gave us any of them, but kept them to sell. One day he had about thirty or forty pint jars full on a table. My brother George and I decided we would just swipe a jar, he wouldn't miss just one pint (we thought), but he had them all counted. While he was out one day oystering was our chance. We took a pint, and I am telling you, they were so good. George always loved oysters better than anything he ever ate. When Grandpa counted them and one jar was missing he told Mama that one of his jars was missing and he had a pretty good idea where it went. That night Mama called me and George in and we had court and were tried and found guilty and were convicted and believe me, we paid a penalty we didn't forget— a good whipping. But it was worth it the way those oysters tasted.

Grandpa Stephens took a mail order bride from Massachusetts, and the poor woman thought she was getting an ideal husband and he thought he was getting a rich woman.[4] Both were disappointed, and the woman took sick and stayed with my parents. My sister Winnie remembers they gave her the front bedroom in the house and our mama took meals to her room. Since there was no indoor toilet she had to use a slop jar and would pay Winnie to empty it. Her family came and got her.

My dad supplied Grandpa Stephens with Carter's Liver Pills. On some occasions my dad would roll up flour and water into pill shape and give it to the man, who took it and quickly felt better.

In a small two-story house down near the river lived an old Norwegian man, George Christian. He took care of our chickens. He would yell "chick-o chick-o" every afternoon at exactly four o'clock and would not feed one until all were there. The chickens knew his call. They would come from every direction. He was also janitor at the church and kept it in good shape—clean and spotless and the kerosene lamps filled and trimmed. He was always present when any church meeting was going on. He gathered oysters in season for the Key West market. He would go out about two times a week. He'd gather them and put them in a pile near his dock in water—water kept them fresh. The day before my dad was to sail, Mr. Christian would sack them up in corn sacks for Papa to take them to Key West. No one ever bothered his oysters. When he wasn't oystering he made cast nets in all sizes of mesh and twine and lengths. We had lots of them. He also kept buttonwood for the wood stove, cut with a bucksaw. I would file the saw for him, and that was how I learned how to sharpen a saw.

I EAT ANY THING, FRESH OR STALE. BUT I HAVE GOOD MEAT IN MY CLAWS.

BLUE CRAB, GOOD SALE IN THE MARKET

Blue crab.

"HE SAID HE HAD BEEN PATTED ON THE HEAD BY NAPOLEON." ROB

When I was very small I can remember Mr. and Mrs. John Gomez and just how they looked—their faces dry, brown, and wrinkled. They visited us and stayed a day or two. They both smoked a clay pipe with a reed stem. They had a buckskin bag. I can remember they called it the "backey bag," and when they emptied out their pipe of burnt tobacco they would empty it back into the bag with the other tobacco. I even remember how bad it smelled when they lit a new charge. Gomez lived on Panther Key or Gomez Key and was said to be about 120 years old when he died. Gomez was a very religious man and interesting. He said he had been patted on the head by Napoleon. He told my dad that he was born in a French province and had proof of his age. He had been married twice and had a son by his first wife. His second wife, Sarah [from Georgia], who lived with him on the island, was about ninety years when she died in Fort Myers about four years after her husband's death.

Gomez joined the Methodist church at Everglade at the age of 115.[5] He had been a Catholic. The minister set aside the last Sunday in any month having five as Gomez Sunday and the entire congregation met at Gomez Point, where services were held. My dad transported the congregation to the point on those occasions.

One Friday afternoon in 1901 Gomez went fishing in a rowboat and did not return. His body was found hanging by his trousers from a limb of a tree which was overhanging the water. It is believed the strong current swept his boat under the tree, and after his clothing became entangled in the limbs the boat was swept out from under him and he drowned.

Gomez was buried on the point. His wife went home with our family after the funeral.

My dad said Gomez told him he was the nephew of the pirate Gasparilla.[6] He lived a very poor man, and Sam Williams [a Marco boatbuilder] teased him about his treasures—saying why didn't he go dig up some of his uncle's treasure. Gomez said there was buried gold up Rice River, but he wanted no part of the blood money. Sam told him to tell him where it was, that he would go get some of it. Gomez said it was up a big river east of Boca Grande. I suppose it must have been the Peace River, for some treasure has been found up there. I've been told of charts that show more treasure up there, and I saw one chart, but no one can read it. It was found up there. And I have a copy of another chart, but cannot read it. I helped hunt for some of the gold in later years that was supposedly buried up the Peace River. Gomez said he climbed up on the mast of his uncle's ship and watched the crew take several boatloads of gold up the river, where they buried it. Charley Johnson and I hunted for it but found nothing, even though our machine registered something beneath us, but it was at the side of a muck pond and had settled too deep to recover. We found plenty of poison ivy. Up the river, gold has been found.

Even though I was only seven years old, I remember the day Gomez's body was found. It was about twenty-four hours after he had gone fishing. There was an island near Gomez Point called Three Brothers, and the brothers living there at the time were the first ones to find Gomez after he was missing. His body was found with his wrist still tied around his cast net. Crabs had eaten his nose and ears; the soles of his feet had come off. Rumors circulated that he had committed suicide, but my family always thought it was accidental. Mrs. Gomez had told someone that her husband had said he was tired of living, but he was never seen without his dirty white canvas cap and clay pipe—these were found on the boat seat. She wanted to go look for him but was sick and didn't. The secret to a long and peaceful happy life is to honor your father and mother, and I have done that.

"WE WOULD PICK FLOWERS, AND ONE DAY MY SISTER, ELLEN, GOT TOO CLOSE TO A BIG MOCCASIN AND HE BIT HER ON THE LEG." MARILEA

I was born and grew up near a little town near Tampa, Florida, in a small town called Abbott, now called Zephyrhills. We lived there until I was about five years old. I was the youngest of four children, three girls, one boy.

My father decided he wanted to travel to see the big cities, so we moved to a little place called Bloomingdale. My father had six brothers. One lived down near Venus, Florida, and wanted to move up near where the other brothers were. Three of the brothers made plans to move him. All the families went along for the ride. There were five covered wagons. Sometimes the older children would walk with the men behind the wagon. We would pick flowers, and one day my sister, Ellen got too close to a big water moccasin snake and he bit her on the leg. I don't know what they did for her. I can't remember any medicine we ever had but castor oil—maybe they gave her some of that. I don't believe they prayed or I would have remembered that. They didn't know how to pray.

Papa became a hardshell Baptist. I never knew what that shell stood for, but one thing I did know, it was hard. But he went to church later on and saw the light and was a born-again Christian. He lived a Christian life after that. He was over a hundred when he died in Naples. My grandfather lived in Florida all his life, fought in the Civil War, and when he came back he got a pension. He was paid in gold every month. Before he died he made sure my sister Ellen, his favorite grandchild, got all the gold that he had saved up. She was the only grandchild he paid any attention to.

One day we visited a cousin in Springhead [near Plant City], and she was wearing a loose dress and got too close to a fire and when it caught her dress on fire she burned to death. This was very sad. I had another relative there that got struck by lightning and died as he stood under a big oak tree.

My mama, Mary Chancey Pollock, was a strong southern woman from Georgia whose first husband ran off with another woman and was never heard from again. She grew cotton and plowed until she met and married my daddy. We were raised poor, although we never went hungry. We didn't have things. My father was a hardworking man, clearing land by hand with a grubbing hoe. I never remember celebrating Christmas, except once. It was just another day for us. We always went to church, though, but we were so poor there were no trees or gifts. One year I recall Papa dressing up like Santa and we played games. Everyone had a good time and never forgot it.

One thing we always had plenty of was running water. All we had to do was run to the pump, pour a little water in it, and pump with all our might. Sometimes Papa would have to find an old shoe, cut a side from it, and put it in the pump to make it work. We had three washtubs and one iron pot to do our wash. We boiled our clothes. I don't know why, but everybody did. To make the water soapy we would cut up Octagon soap in the pot, keep pushing the clothes down as they boiled, and after a few minutes take them out of the pot,

put them in a tub with clean water, then soap again and run on the washboard again. Now they were ready to rinse in two clean tubs of water, wring out, hang on the fence, pray for it not to rain so they would dry before night. The next day we would build a fire in the stove, put three irons on it—maybe put on some beans in an iron pot, sweet potatoes in the oven—then go outside, sit in the shade where it was cool, until the irons were hot and the house hot. We were ready to iron just a few things.

We didn't take a bath very often—sometimes on Saturday night—but one thing for sure, we had to always wash our feet with soap. It was a lot of trouble to take a bath in a washtub. We never had any detergent of any

kind—all we had to wash with was brown Octagon soap. It did the work, I guess, because we all smelled the same and no one complained at church.

Most all the women wore what they called "Mother Hubbard" [a long loose-fitting dress with gathered apron]. We lived in Arcadia until I was sixteen, and these women never changed their dress style, so you never knew when they were going to bring a new baby to church. The women knew what was causing them to have a baby every year, but they didn't know how to stop it.[7] There was only one doctor in Arcadia, and he wouldn't have known what to tell you. He probably would say, "Take an aspirin and call me in the morning."

"ROB WENT OUTSIDE AND WRAPPED HIS ARMS AROUND A PALM TREE AND HUGGED IT FOR A LONG TIME." MARILEA

My father-in-law, R. B., died early in the evening. My sister-in-law Dolly [Rob's brother Hub's wife] sat with me with him. He had been sick for three days and a nurse had come to help. The nurse took his temperature just after he stopped breathing. She did it out of curiosity, to see how hot he really was, and it was 106 degrees. His breathing was strange and would stop and start. It sounded like an old wringer-type washing machine. I couldn't stand to wash clothes after that. Rob put his hand over his papa's chest and felt his heart

quiver, then stop. Winnie [Rob's sister] came in and got in bed with him and put her arms around him and cried. Rob went outside and wrapped his arms around a palm tree and hugged it for a long time.

Nancy lived twelve years after my father-in-law died. She loved cats and had one with two different-colored eyes and a bobtail. One day on her way to a prayer meeting she slipped and fell on a wet stepping stone. She broke her ankle. Four days later she died at Winnie's house. Winnie called Rob when she saw Nancy acting strange. When Rob and I

got to her bedside she whispered to Rob, "Rob, pray, I'm going." Rob cried, "Oh no, Mama!" She said, "Yes, I'm going, I'm gone," and she died. I closed her eyes and mouth. I remember seeing an awful moment of fear in her eyes and that scared me. Rob walked the floor rejoicing, while Winnie cried something awful. He was thinking of the verse that says to rejoice at death and weep at birth. I remember it was a strange afternoon—the sky was a eerie yellow color, a sign that a thunderstorm was passing.

"WHEN I CRAWLED OUT ON THE BANK THE GATOR WAS ALMOST CLOSE ENOUGH TO GRAB ME." ROB

We swam in the river almost every day. It is a wonder we didn't get gator-caught. One time a gator caught one of the our dogs and killed it. We could always catch a mess of sheepshead and snapper off the riverbank. We would play hide-and-seek in the cane fields behind the house and would hang our feet in the water and let the catfish suck our toes. Behind the woods in back of the house was a prairie and always lots of ducks and curlews. We had fresh meat almost any time we wanted it, and that was often.

My brother and I were swimming one day when I was about fourteen. My older brother Claude and I had just unloaded a skiffload of buttonwood and decided to go in swimming. There had been a big gator in the river, and several people had shot at him and thought they had killed him because he sank. I said to Claude that I was afraid of that big ole gator, but he said someone had killed it. I said I didn't know that for sure. We went in and were almost out in the middle of the river, swimming. Claude thought he would have a little fun and fool me. He started swimming fast and yelled, "There comes that gator!" I noticed he was laughing and didn't seem very excited about it. He said he was just fooling me. I told him he ought not to be fooling that way. But I looked and there he was, really coming fast. I yelled, "There he is sure enough!" And it was a race to the riverbank. When I crawled out on the bank the gator was almost close enough to grab me. I ran to the house and got the shotgun and a handful of shells. But they were fine shot.[8] I shot the gator and put both his eyes out and he swam away. The next day Raleigh Wiggins found him sleeping on a mudflat and killed him. He was thirteen feet and five inches long—the biggest one I have ever seen in my whole life or ever heard of. He was the kind that would catch you.

"WHILE I LIVED ON THIS OYSTER BAR, SOMETIMES IT WOULD BE A WEEK AND YOU WOULDN'T SEE A BOAT PASS." ROB

My first trip to Naples was when I was a boy, when my papa and I visited the Russell family at Rock Creek and we had to sleep in the barn. We walked up the beach, and the Russells took us to their house by boat. We crawled through a hole in the ceiling to reach a small upstairs loft where they kept hay. I remember curling up close to my papa because it was so cold. He said he hoped the barn didn't catch on fire because we'd have a time getting down. I couldn't sleep for worrying about a fire. The Gastons lived on the west side of Russell Island, the Russells on the east. The Russells lost a young boy when I was a boy. Harry McGill and I dove for the drowned boy, but my dad hooked onto the clothes with the grappling hook several hours after he drowned. The Gastons raised beautiful chickens and sold eggs for thirty cents a dozen.

In 1916, just before my marriage [to Cassie], I had been to Bonita Springs to church, and on my way back to Chokoloskee I went into Gordon's Pass at Naples and saw how beautiful it was and that the fish were plentiful. I decided to live there and tried to talk my brother George into moving, but he wouldn't budge. I talked Mack Johnson into moving and being my fishing partner. He rented an old cane scow to move. I rented one of George's boats.

The next year George came and liked what he saw and moved. The South Fish Company from Fort Myers furnished us with a small boat with an icebox, and we hauled fish to Marco, where a run boat would come twice a week to get our catch. Our supplies came from Captain Bill Collier's general store in Marco, as Naples had no store.

EVERGLADE

CLAUDE
AND
ROB

A NARROW ESCAPE
WE ALMOST GOT CAUGHT

MY brother and I
were swimming one
day. we saw A BIG GATOR
coming, we gave him a
good race, "WE WON", But
it wasent any thing to
Brag About. when we
Reached the Shore he was
in a fiew feet of us.
I ran to the house and
got the shot gun an give
Him 3 Loads of shot befoy
he would leave, I put Both Eyes
out and he desided he had Better
leave. The next day Raleigh Wiggins
killed him. He was 13 feet 5 inches
& the Biggest one I have
ever seen in my whole life.
or ever heard of. He was the
kind that would catch you.

JUST A LITTLE SKETCH OF OUR
HOME AT EVERGLADE WHEN I WAS
A BOY. I WAS BORN IN THIS
HOUSE. MAMA & PAPA RAISED
NINE CHILDREN IN THIS HOUSE.
WE SWAM IN THIS RIVER
MOST EVERY DAY. IT'S A WONDER
WE DIDENT GET GATOR
CAUGHT. WE COULD ALWAYS
CATCH A MESS OF SHEEPHEAD
AND SNPPER OFF THE RIVER
BANK. BEHIND the WOODS
BACK OF THE HOUSE WAS
A PRARIE AND ALWAYS LOTS
OF DUCKS & CURLEW - WE HAD
FRESH MAAT MOST ANY TIME
WE WANTED IT AND THAT
WAS OFTEN. THOSE
MEMORIES WILL NEVER
BE FORGOTTON. THE
YOUNGER GENERATION
WILL NEVER KNOW
WHAT THIS KIND OF
A LIFE WAS.
ONLY THE MEMORIES
REMAIN JUST A
FIEW LEFT TO
TELL THE
STORY.

2

Almost caught by a gator.

My First Home at Naples

This is where George and I lived when we first came to Naples 1916

Located in Naples Bay – in front of Jackso Fish Camp where is now. 1917

Looking up Naples Bay.

My first home in Naples was on an oyster bar in the lower end of Naples Bay. George and I picked the spot because we could get a good view of the pass and the breezes were excellent. Schools of mullet had to come by there before going out in the gulf. We cut black mangrove and buttonwood trees and worked them down into the mud for pilings. George watched the Indians make cabbage fan roofs and decided we'd use that for our top.

While the house was being built Cassie and I moved from a tent into a vacant house in the cove at Jackson's Fish Camp, belonging to Charley Johnson. We had dug a well and curbed it with a barrel. It had lots of iron in it, but it was good drinking water. We had to carry our water about three hundred yards.

My first home in Naples was a wonderful place to live at that time. Fishing was good, even at the door. Oysters were good everywhere; we had no pollution. There were plenty of ducks, and all kinds of good eating birds. Plenty of deer close by, easy to get. While I lived on this oyster bar, sometimes it would be a week and you wouldn't see a boat pass, but now there is one a minute, the whole day long. Now there are million-dollar homes on this spot.

Cassie and I had a fifteen-year-old boy, Mac Hicks, that stayed with us. He was a good boy and loved to wash dishes. Cassie never had to wash them. He'd wash and rewash them and sing. Once he stuck a splinter under his foot skipping and dancing on the rough lumber floor. He asked me to pray for him. I convinced him it had to be cut out, though I didn't know how. I took a razor and pliers and got it out. Later, he died with the flu, the same year a neighbor, Henry Daniels, lost his fourteen-year-old girl to the flu. Prayer was about all we knew to do. When George and I were putting the roof on top of the little house we bought, George got the flu. Just as I got the tar

Rob as a young man, just before he married Nora Cassie Williams.

paper on I came down with it. Brother Babe and Sister Sally [the preacher and his wife at Chokoloskee] put a Bible on my chest and prayed.

"HE GOT A LITTLE CLOSER TO MY SIDE OF THE ROAD AND IN A LOW VOICE HE ASKED ME TO MARRY HIM." MARILEA

One day two cars came up to our house. Three men and one woman were on their way to a camp meeting [religious revival] up near Tampa. My fisherman [Rob] was one of the men in the car. He got a chance and asked if I would like to go with them. I told him Papa wouldn't think of letting me go, so don't dare ask him. Well, they didn't stay but a couple of days. We were going to prayer meeting night; it was about two blocks from where we lived, so we just walked. Papa asked them if they would like to go. They said yes, so we all walked over together, but coming home later after the service was over we got separated and he got on one side of the road and me on the other side. Papa and my uncle were right behind us, but before we got home, he got a little closer to my side of the road and in a low voice he asked me to marry him and go to Naples with him the next day. I told him that was short notice. I knew I was going to do it—still, I wanted him to think I was hard to get, and I didn't know what Papa was going to say to me, being the last one at home. My brother and two sisters had already married. The last one had run away, but I didn't think this fellow would steal me, because he was a Christian. I told him I would think about it and let him know in the morning. I went to bed early, but not to sleep.

When he came the next morning I had my suitcase packed. I told him I would go. It took him about two or three hours to get up the nerve to ask Papa. He finally got nerve enough to ask, and I don't think Papa ever told him yes, but just how hard it was to control me. I guess he thought if he told him all that he wouldn't take a chance on that for a wife. Anyway, they wound it up some way and I went in, put on my white dress, got my suitcase, and told Mama goodbye. The only thing she said to me was, "Baby, be sure and clean his clothes when you wash." She lived about two years after I married. I visited her often and would go up and stay a week with her.

After I put on my white dress and was ready to go, we drove to the courthouse in Arcadia and were married at ten o'clock in the morning. Papa and his brother went along. Then we headed for home, wherever that was. I was a dumb bride. I didn't know where we were going—to Naples or Everglade—and didn't ask anything about where we would be living. I trusted this man I had just married. Every-thing I owned was in one suitcase. Rob had a suitcase full of dirty clothes.

We got to Naples about dark after traveling all day. There was a lady on Twelfth Street that had a large house and she rented rooms. It was the only place you could get a place to sleep in the summer. The Naples Hotel was built then, but it closed for the summer every year. And come to think of it, it took a lot of money to stay there. Early the next morning Rob told me that we were going to the bay and our home, and I didn't even know what a bay was, but I got my things and we went out on the dock and he showed me his boat, the *Seabreeze*. It was tied up to the dock, and I had never been in a boat before. When I got in the boat there were dock fiddlers that came running out, and I had never seen these before and screamed. Rob said they wouldn't hurt me. But I didn't believe him.

When we got down to the house I got a surprise. The house was built on an oyster bar high up on stilts and was made of pine boards. The windows were the same. You just shut them and fastened them with a string. There wasn't anything to keep the mosquitoes from taking you away. No use to shut doors and windows because they came through the cracks in the walls and floor. Everybody had to make mosquito nets, but in case you turned over in the night and stopped against the bar you didn't stay there long. The top of the house was covered with [sabal] palm fronds.[9] When the tide would rise it would cover the oyster bar. I thought sometime it might come in the house, but it never did until we had a hurricane.

One of my husband's brothers [George] and his wife, Lanora, lived in one side of the house. The kitchen was in the front room. We had a two-burner camp cookstove that burned gasoline. A blanket was nailed up for a door separating our sleeping rooms.[10] When they built the house they made a little john in the back and a walk to it, but it was full of fiddlers. When Rob was there I would get him to go with me to run them out.

We didn't have much to cook—fish, rice, tomato gravy—and didn't have an oven. But we solved the problem by making what was called a hoe john—cook it on top of the stove. Sometimes it was burned and sometimes it wasn't cooked enough. We just ate the good part.

I didn't know fishermen fished at night, so the first time Rob said he was going fishing a little while and wouldn't be gone long I was scared. I was scared of the Indians. I thought this may not be so much fun after all. I wondered if I should have stayed with Mother. So the only thing to do was go fishing, so I went. I didn't know there were so many mosquitoes in the world. At the oyster bar house Rob would cut dead mangrove wood, bring it in, make a smoke, and run some of the bugs away, but they came right back and brought more with them.

We all wanted to move from the oyster bar. A year or so after we married the boys tore the house down and we moved up near the bay dock. Everyone that owned a boat had it tied to the dock. There was an old house there where they opened oysters. There were houseboats. Rob's daddy and one of his brothers built boats there. The mosquitoes were so bad because the house was made of boards that the mosquitoes could come through. We made a mosquito bar and would crawl under it at night and tuck it under the mattress, but it was a problem with the children. We had no water, only rainwater, and I had to wait until it rained to do wash. Later, when we had a pump, the water had so much iron in it that we had to boil it and let the iron sink to the bottom. I remember my mama soaking nails in water, and when they got rusty she made my sister Lottie, who was sickly, drink the water. Another woman I knew would file off the rust from her garden hoe, sift it, and make the kids drink it.[11]

In the winter the hotel would open and people would come down to fish and get some sunshine. Rob and some of the other fisherman decided to try guiding. They didn't have any trouble getting parties. They would go to the dock and wait for jobs. Sometimes they would go in the evening to the Naples Hotel and try and get a party for the next day. The party furnished Rob's lunch and paid him three dollars a day from eight in the morning to five in the evening. That was good money. Fish were plentiful. It was not unusual for Rob and his fishing party to hook twenty-five tarpons in one day. Rob enjoyed it as much as the party.

An old lighter that was turned into a house boat. Pappy Turner's *North Star,* part of the early Naples Mackerel Fleet, is in the foreground.

"ADNIE AND I USED TO GO WHERE THEY DUMPED RIPE TOMATOES AND GO FIND COKE BOTTLES, MAKE KETCHUP AND CAN TOMATOES." MARILEA

Rob and Forrest Walker fished together for many years. We had four children, they had five. One of his boys became Speaker of the Florida House of Representatives. Adnie and I never had any trouble, but one time we had a few words. It was all over in five minutes and forgotten. She was the best friend I ever had. We built a fence between their house and ours. We would sit for hours and talk. The kids soon tore the fence down, then it was easier to get to our back door to borrow sugar and tea. The kids had their fights, pushing each other in the railroad ditch, which was always full of water in the summer, and Adnie pulled one of the kids out with a rake once to keep him from drowning. When it would rain the kids would play naked and get a good shower. Adnie and I used to go where they dumped ripe tomatoes and go find Coke bottles, make ketchup and can tomatoes. Before we got a refrigerator we had to buy ten cents' worth of ice every other day; sometimes I had to borrow the dime from Adnie, and sometimes she borrowed one from me. Most of my life had been spent in isolation from close neighbors.

We survived on fish, cotton-tailed rabbits, and chickens. Sometimes the men would kill a deer and we would have plenty of meat. Sometimes if Rob had a good week fishing I would go up to Bonita Springs and buy a carload of groceries. I remember salt pork was nine and ten cents a pound. I would load up that Ford with groceries for ten dollars. There just wasn't much to buy.

"SHE CHARGED TEN DOLLARS TO DELIVER A BABY." MARILEA

Our first two children were born at my papa's home in Arcadia; Olivia was born at R. B.'s house in Fort Myers. They had moved there so Winnie could go to school to be a teacher. Just before the depression our fourth child was born. As I looked at the tiny twisted body I said she couldn't be mine, I had three healthy children.

When I became pregnant in the seventh year of my marriage the town had no doctor in the summer months. The nearest one was thirty miles away, and in those days it would take two hours with good luck to go that thirty miles. The roads were made of oyster shells and some conch and you always got at least two flats on the way. The roads were like a washboard and you had to drive real slow or you might lose a fender.

I started having pains about ten days before the baby came. There was a midwife here, Lizzie Bostick, and she delivered all the babies. She charged ten dollars to deliver a baby. My husband had her come over. She had me stay in bed and said I would be all right. How can a woman with three small children stay in bed two months? So I would get up and help, and in a day or two I went into labor again. My husband had the lady come over again. She said we had better send for a doctor. She had never had a case like this. A friend of mine, Minnie Townsend, who lived at the City Dock on a houseboat, came over and stayed with me. She had a little girl about two years old that she left in the care of an older sister. Late in the evening the police chief, Cale Jones, came out and told us that the little girl had fallen overboard and drowned. That upset everyone. Minnie left immediately to take the body to Zolfo Springs for burial. On the way they stopped in Fort Myers to contact Dr. Winkler about my condition. He sent some pills back with them, but they had no effect.

We had one of the first phones in Naples and we called Fort Myers. It cost us thirty cents. Dr. Winkler sent Dr. Johnson down about midnight. The baby came about five in the morning. He charged seventy dollars to deliver her. I knew something was wrong with her because they wouldn't let me see her. They told me she weighed three and a quarter pounds. They took her away across the hall to another room. All morning friends came and went in to see her, and I was sure by now that something really bad was wrong with her. I asked my husband to bring her to me, but he said, "No, let the midwife take care of her." He said I should rest. Later in the evening he finally told me that the doctor said she couldn't live but a few hours. I said if you don't bring her to me I will get up and get her myself. He finally brought her to me. What a shock I got. She looked more like a monkey than a baby, only she didn't have any hair on her body.

Oh, the things that went through my mind. I didn't have anything that would fit her, not even a diaper. So we would just put oil on her and wrap her in a cloth. Each time I saw her body I would say she couldn't live. There wasn't any flesh on her. It was just skin over bones. But she continued to live. I had plenty of milk, but she wasn't able to nurse. Every hour I would get a tiny spoonful and get what I could down her. Lots of times I would choke her as I tried to feed her and I would think, She's gone now.

A good friend, Susie Addison Whidden, came to help me. When she left for home she found Enid, her husband, had shot himself. His body was kept in the ice house for a few days.

I can look back now and it had to be the hand of the Lord that kept her alive. We had never heard of such a thing as an incubator. I found out years later that they didn't have one in town where the doctor came from, so we just carried on from day to day trying to keep her alive.

I had a lady come in and stay to do the housework and help with the other children so I could put all my time in caring for the baby. She began to gain about an ounce a week. Then I began to have hope. My friends got

together and made her some dresses from a doll dress pattern.

When Betty was about six months old she had long dark hair and a pretty little face. But still you couldn't look at her body and have any hope. When she was just born she made a noise like a little puppy. She never cried until she was two months old. When she finally did cry I was sure she would be all right. But when she was a year old my mother-in-law said she wanted me to take the baby to a specialist, that she believed there was something wrong with her, she didn't hold her head up and didn't crawl. It hadn't even crossed my mind that there was anything wrong now. She was just little and would be all right as she grew older. But we took her to the nearest specialist and he looked at her and said what

Rob and Marilea with their four children (*left to right*): Lemuel, Betty, Olivia, and Marguerite—all born before there was a hospital in Naples.

we had hoped he would: She was just little, but would be all right. We came home happy to announce to our friends that she would be all right, but as she got older I would notice people staring at her when they didn't think I was looking.

But our life went on. The strange thing about these children is that everyone notices it before you do. I guess deep down you know, but you just can't face it. But there comes a time when you have to face facts.

There was a doctor that had a winter home here. When Betty was about three years old he came to spend the winter. Betty got really sick while he was here, and we asked him if he would see her and he did. He treated her and later told us that there was a doctor coming to St. Petersburg to see these children up there and he wanted us to take her to see him. So we did, but when he looked at her he said he could have filled a hospital in three days with children just like her. We didn't know there was another one like her anywhere. He said the best thing for us to do was put her in a home; that he didn't think it was right for a young couple to be tied down with something like that. He said he would make all arrangements to admit her, that she was a cerebral palsy child and wouldn't ever get any better. Rob and I didn't like what we heard and told him so. Since that time we have carried her to every doctor that we have heard about that has had any experience with these children, but no results.[12]

I think the hardest part was when she was old enough to start talking and walking and my friends' babies were doing all the things a baby should. That was when I had to face things just as they were. I got to the place where I could no longer expect a miracle. I was struck by the reality of her condition and torn inside. I have many friends close to me who share my problem, but they never really know what it is like to have a baby like Betty. They are sympathetic, but removed from the real facts. She didn't say Mama or Daddy until she was six years old. There is never a morning when I go in her room to get her ready for breakfast that I first don't look to see if she is breathing.

"YOU JUST DIDN'T LEAVE SOMEONE ALONE TO DIE." MARILEA

I remember taking sandwiches to the house of the dying, waiting, sitting, resigned to the reality. It was a custom for the women to sit with the dying all night. We would put a little Vicks VapoRub around our eyes to keep us awake so we could listen for signs of life. You just didn't leave someone alone to die. There were four of us: Adnie Walker, Zola Summerall, Flossie Pollack, and me.

I sat with Mr. Driggers at the Haldeman place. He was the father-in-law of Fred Williams, the Haldeman caretaker. The men bathed Mr. Driggers after he died and dressed him for burial.

After Irma Rigsby's mother died we bathed and dressed her in clothes the family had put out. We put pennies on each eye to keep them closed. The men brought us a board which was covered with a sheet and called the cooling board. Sometimes a door would be taken off its hinges and used. I remember Irma had been sleeping in the same room when her mother died. She jumped up and ran over and said she wanted to feel her mother's pillow. She grabbed the pillow and said she felt one lump of feathers. She believed if the feathers formed lumps it meant she'd gone to heaven. We turned the body over and heard a noise and it scared us so bad we ran out of the room.

We always sat in twos. Adnie Walker said to me one day she wanted me to promise that I'd sit with her when she died. Nobody was embalmed. People were buried at Pioneer Cemetery or Rosemary Heights Cemetery. When Noah and Jessie Kirkland's daughter died my kids picked wildflowers and put them on the grave. Now there are just weeds there. There are several suicides buried there. One was a mother of several small children who committed suicide on Christmas Day. There are several who died of abortions—one woman aborted her baby with tweezers and died. Another was married to an older man who was cruel and when she got pregnant with her third baby she could not go on. After she died, one of her daughters bore her own father's children. They moved from Naples.[13] Robert Davis and Morris Bostick were buried there, killed in a motorcycle accident on Fifth Avenue while riding in the dark.

"SOMETIMES A HURRICANE WOULD RUN US BACK HOME, SOMETIMES WE STAYED IN A TENT." MARILEA

Each summer we would go down the coast to Chokoloskee. Fishing was better down there. We would stay down there until about November, when the fish would start running at Naples. Sometimes a hurricane would run us back home, sometimes we stayed in a tent. Later, Rob built a houseboat. By being pushed in the railroad ditch, my children could swim by the time they could walk good, so I didn't worry about them falling overboard. Soon the children were old enough to start to school, then I had to stay home and put them in school and Forrest and Rob would go for a week's fishing. We had a cabin boat. They could eat and sleep on her and fish from the little boats. When the boys would come home it was like having a second honeymoon.

I used a rub board to rub the clothes when I washed—that was the way poor people washed. One day I saw in Sears Roebuck that you could get a washing machine run by a Briggs and Stratton motor—on credit, so much a month. We bought one. Adnie, Forrest's wife, heard the good news and bought herself one. We wondered what would hap-

Fishing at Burned Stump, where a mangrove tree was once burned.

pen, since we stopped boiling our clothes, but it didn't seem to make any difference.

By this time our oldest boy was in school. That was a sad day for me. Soon the other two were ready for school. I had three in school, and the Walkers had three. They would walk across the railroad, where the bus would pick them up. My youngest girl, Olivia, was always the first one to get to the bus, and in twelve years she wasn't absent or tardy. You would hear the screen door shut, and two or three minutes more you'd hear it again. The bus driver wouldn't dare leave until all six were aboard.

There was no lunchroom at school, so I would pay my sister Lottie, who lived near the bay, to give the kids lunch. They ate great northern beans and biscuits every day. Later, they went to Dyches Store and got banana and mayonnaise sandwiches. Eventually, a little building was set up next to the school by someone to serve lunch, but they only served soup.

The oldest Walker boy and my son and girls finished high school. My husband, being a fisherman and knowing all the hardships, and I didn't want our son to be one, so we sent him to Jacksonville Barber School. He finished, came home, and went to fishing. The day before he left for Jacksonville he married Mr. McKinney's granddaughter, Dan House's daughter Lucy. We didn't know he was married for three months.

SOMETHING I CANT FORGET

1913 - A BAPTIZENG AT CHOKOLOSKE ISLAND
AT MR. MCKINNEYS LANDING. MINISTER BRO., TOM CREWS
40 WAS BAPTISED, A VERY PRECIOUS MEMORY
OF THIS DAY. I WILL NEVER FORGET.
SOME WERE LAUGHING, SOME WERE CRYING SOME WERE
SHOUTING AND PRAISING GOD + REJOICING WITH HPPINESS.
WHAT A DAY - IT SEEMED THAT HEAVEN OPENED UP.

MR MCKINNEYS HOME
AND STORE

Chokoloskee Island, where forty were baptized.

FOUR 🐟 FAITH AND RELIGION

"FOR WEEKS THEY LIVED ON WILD FIGS FROM A HUGE FIG TREE IN THE SCHOOLYARD AND DRANK FROM THE SCHOOL'S CISTERN." ROB

Following a revival in 1913, forty of us were baptized at a service at McKinney's Landing.[1] Tom Crews conducted the baptism. Mr. McKinney remarked afterward that he wished they would have done it somewhere else, as he didn't want all those sins washed up on his yard. Even Tanty Bogus Jenkins, the old alligator hunter, and Harvey Daniels were baptized that day. Ida Lopez was too, and I toted twenty-five roses twelve miles from Deep Lake to Everglade in a shoebox for her once.

A Pentecostal religious outbreak spread swiftly in the area just after the 1910 hurricane when Jim Addison, a fisherman, and his wife came back from California with a new gospel. He became involved with a group that was teaching the return of Christ. They started holding services wherever they went. I first heard of the new gospel through Erskine Roll or Pat, who was living with my uncle. As the doctrine spread, so did repentance. Overnight, many professed change and gave an honest effort to try to do good. It was drawing people together. Chokoloskee was the last

of the islands to give into the outbreak. Smallwood called it the "tongue talking religion."[2]

In 1912 Jehu "Babe" and Sally Whidden, along with Tom Crews, caught a boat from Fort Myers to Chokoloskee and brought their gospel with them. The spirit directed them to go to Fort Myers, where they'd find a boat waiting for them. They found Willie Brown there with a sloop. He had just been baptized on the clam bar at Marco. He took them to Chokoloskee.

These missionaries, originally from Venus, Florida, were at first met with opposition. They arrived penniless and found refuge in the small schoolhouse. For weeks they lived on wild figs from a huge fig tree in the schoolyard and drank from the school's cistern. As the gospel from Everglade spread, the missionaries gained acceptance at Chokoloskee. Grandma McKinney was the first convert. She and Grandpa McKinney got married, since they weren't married. She would dance and shout at church meetings.

I remember Harvey Daniels when he was just a young fisherman. He loved to drink. When he couldn't get whiskey he would go

to the store and buy all the vanilla extract he could or anything else that had alcohol in it. This would happen almost every Sunday. When the revival broke out Harvey was converted, and instead of getting drunk every Sunday he was in church and was a different person for a while. But soon he backslid and was as bad as ever. One day he passed by a gasoline drum that had been emptied on the dock at his father's place. Just to see what would happen he lit a match and threw it in front of the hole where the fumes were coming out. The drum burst and shot flames out and burned him and blew him overboard and almost killed him. He was burned so bad you couldn't tell who he was, and his face was so black and swelled it was really an awful sight.

All his folks were praying people. For several Sundays we gathered at his home for church service and to pray for him. He got well and came back to church for a while, but it did not last. One night he and Uncle Dick Myers were coon hunting near Dismal Key. He was hunting in the Dismal Key Cove. He shot a coon and got out to get him and stumbled over a mangrove root and stuck his shotgun barrel in the mud, and when he got back in his boat he just thought he would hold his gun up over his head and shoot the mud out of it. It was a Winchester shotgun, and when he fired it the gun burst and some of the works went in his forehead and shook him up real good and all he could do was go around in a circle. Uncle Dick was just around the island and he heard him shoot, but he knew something was wrong because there was lots of coons and he had stopped shooting. When he got to him he was just going around in a circle. They took him to Arcadia Hospital, as that was the best hospital in the area at that time. They operated on him. Dr. McSwain got some of the iron or steel out but overlooked a piece of it, and soon Harvey got worse and died.

I remember the revival meetings in the schoolhouse at Chokoloskee well. Old Mr. McKinney would go out to hear the preacher preach, but he wouldn't take any part in it. He'd only sit and fan as hard as he could with a turkey wing. He had a little newspaper and always included news about the new preacher and about the "hellcats" or local bad boys. He said he always knew when they were going to make moonshine because they would come to the store and buy a sack of corn. Sometimes drunk young men would come to church and sit on the back seat and try to interrupt the service by laughing and talking loud. We ignored them.

"SHE CAME TO THE STORTER HOME AND LOOKED LIKE A WILD WOMAN." ROB

One night in church Babe and Sally's daughter, Ava, a fourteen-year-old, stood up and testified, and Charley [Johnson] got up and ran to the altar, looked back, and said to his wife, "Lillie, pray." Sally and Babe had tried to get Charley to church, but he always gave excuses, like he couldn't lug the baby there. Babe and Sally were without food, and one day Charley saw them eating figs and said, "Hold on there, Lillie, go get one of them ten-dollar gold pieces and give it to them." One day he told Lillie they were going to find out if it was real and they were going to sit right up front. He became a model for the new gospel.

I was good friends with Clarence Brown. He was converted before me. I saw how he had changed, and I wanted to be like him. He said, "Rob, you have to come hear this preacher, Tom Crews, who is holding services in the old schoolhouse." I didn't make it to the first meeting, but the next time I went. I wanted to go up to the altar, but didn't. The next night I went. The two women who were the first converts, Grandma McKinney and Ida Lopez, began praying for me.[3] It wasn't long before I was dancing all over the place—my feet hitting the floor sounded like a Gatling gun. It was midnight when the meeting broke up. Grandma McKinney had an extra room in her house and let the preacher have it.

One time Gertrude Thompson, who once lived at the old Layne place, got it in her head that her husband was going to follow her to Chokoloskee to church, so she disconnected his boat batteries and threw them overboard. She went from house to house prophesying, and people thought she was crazy. She rowed a boat eight miles from her house to Chokoloskee. One night Henry Thompson came in and Bill Gardner was up dancing in the spirit. Thompson said, "I'm going to cut your throat." He was drunk. Bill grabbed him and put his arm around him and kept on dancing. Somebody then went and took the knife from him.

Some kind of bad spirit got hold of Aunt Toggie the same time it did Mrs. Thompson. It happened in church. She got up and started prophesying. She had two of her little boys on a pallet on the floor. She said, "I'm going to be translated [conveyed to heaven without death] tonight, and my two boys with me." The preacher told everybody to get their minds on the Lord. Soon the preacher left and most of the people. Aunt Toggie closed the church door. Clarence Brown and I were outside. Every little while we'd try and peek in, but she'd yell, "I know you're out there." Clarence and I prayed all night. Just before day we left to go home and left her there.

She came to the [George] Storter home and looked like a wild woman. She started prophesying to Aunt Nannie and Aunt Nannie ran her off. She came to our house, ran up the stairs, shut the door, and backed up against it. It took us all to open that door. All the time we were praying, and by this time we knew an evil spirit had gotten hold of her. She finally got quiet and thanked us all for praying for her.

The next day it tried to come back. She lived across the river from us. I heard hollering and went over there. When Aunt Toggie saw me she said, "Rob it's not me this time, it's Bessie [her daughter]." Clarence said, "No, Rob, it's Mama." We began to pray and rebuke this spirit. She was delivered. We prayed and felt it leave. In fact, a force ran us down to the river. I believe that spirit went into the river. We were never bothered anymore. Uncle Mel, Toggie's husband, wouldn't have anything to do with Pentecost. Aunt Nannie and her family were afraid of it. Maybe they were scared off by this experience.

"MAYBE THEY THOUGHT THEY WERE GOING TO THE PROMISED LAND AND THEY WOULDN'T NEED ANY FURNITURE." MARILEA

When my uncle moved to be near my papa, I don't remember anything like stoves, beds, tables, chairs, that my uncle's family took with them. I do remember my aunt, though. She was a little woman, and she had only one tooth right in front. She had twelve children—the last one was a boy. She carried him on her hip all day and worked on (remember, we were living in the Dark Ages then). Maybe they thought they were going to the Promised Land and they wouldn't need any furniture. My aunt was a good woman, but I always wondered why she only had one tooth and no furniture.

One day after we got to Bloomingdale, Papa's nephew came over and told us about a place he wanted Papa to go with him; he said it was a church just a little ways from where we were living. He didn't tell Papa what all happened at this church. Papa told him he would think it over. Papa came in and told Mama about it. She said, "Oh yes, I heard about that place. They tell me the people do all kinds of things, that they put some kind of powder on the people and you can smell it when you go in the tabernacle." She didn't want to go, but to please him she would. "One thing, children," she said, "you sit right beside me where I can see you."

Mama liked the singing, but the rest she didn't care for. Wives obeyed their husbands (or were supposed to). I guess she wanted to see what Papa was going to do. Well, he soon got his hard shell cracked, and Mama saw what it did to him so she started singing. She sang when she was cooking and doing the wash. I was only ten years old, but I thought she was going to get it soon too, and she did. She found out the powder she could smell was pine straw they put on the ground because they didn't have a floor under the tabernacle. Papa soon figured that one out. Papa couldn't stay away. We would go every time they would have a service. It wasn't long before he joined the crowd, then he had to work on Mama. She wasn't easy to change. She liked her coffee, but they told us it was a sin to drink coffee or use tobacco. She would try, and before the morning was over she would get a sick headache. I would be so sorry for her. Most all the people went along with the preacher. Some may have taken a few chews.

When they got home on Sunday night it was time to baptize. Papa said we all should be baptized, so we all went to this little branch [of the river] near the church. I remember it was about knee deep, but we all got wet. We children didn't know what we were doing. We were just following orders from Papa. The preacher said there were forty-seven baptized that day.

Papa had a cousin living in Arcadia, Florida, and we got a letter from her telling us about a place out near Lake Placid where there were about five or six families living—Marion Crews, Mike Bethea, Dave Durrance, and Dylan Whidden. They called it Hen Scratch, about thirty miles across the prairie from Arcadia. She wanted us to come down and go with her out there. It didn't take Papa long to make up his mind. He caught the first train headed south and stayed a week.

He was on fire when he got home. He told us to get ready to move to Hen Scratch. We loaded up and headed south to Arcadia. When we got there they were all ready to go. It was a day's journey with an ox team and wagon. We got there and Papa decided to camp by the side of a little stream that was near the church—a beautiful place—green grass covered the place. We made our beds on the grass, stretched our mosquito nets, and walked up to the church. The building was a post in the ground and small posts on top, sabal palmetto leaves covering the sides and top. The floor was covered with pine straw. The seats were just a board to sit on. We stayed six weeks and went to church night and day. We made meals of sweet potatoes, syrup, and bread on an open fire.

We didn't know anyone there, and Papa wasn't too sure he had found what he was looking for. I noticed when anyone came to speak to Papa they would hug him. And the women would hug and kiss Mama. We had not seen this before (if she was only alive today). She finally said maybe it was all right if they hug and kiss. Every one seemed so happy, and the singing was beautiful. There was one man that had the songs. He would get the song in the wrong key sometimes, but on the next verse he would go up or come down. Then he would sing it over two or three times. You could hear it a mile. They had never seen an organ. No one could have played it anyway. No one there had ever seen a piano. We had never seen such happy people. Papa was so excited. Mama still didn't understand why there was so much kissing, but she finally saw the light.

There wasn't any fresh meat in the store or canned goods on the shelves. In fact, there weren't any shelves in the stores. You couldn't shop around because there was only one grocery store in Arcadia. Shortening came in a five-gallon can. It sold by the pound. They would dip it out and put it on a paper tray. Mama always found something else to put it in before it melted. When we went far we got a greasy spot.

These people were like one family. People camped all around the church. Some had a tent, some, nothing, and when it rained they would pick up their beds and run for the tabernacle. The people that lived there had cattle and hogs. Each family had a big pasture to keep their cattle in, but the hogs were marked. Each family had a different mark. They would kill a hog or cow, bring it over, hang it up under an oak tree, and tell everyone to get what they wanted. They would bring red potatoes, peas, corn syrup, milk, butter, and people didn't eat like they do now. You didn't see any overweight people. You couldn't get fat on what we had to eat, but I never heard anyone complain.

No trouble going to sleep: just roll your bed out on the ground, cover your head so the moon wouldn't shine in your face, and soon you'd be asleep. Papa would build a fire with logs and it would last all night, and if it was cold there would be frost on our beds. If it rained we would have to get up off the ground and roll our beds and get them in the wagon and sit it out. The services would last sometimes till midnight. Sometimes the morning service would run right on into the evening service, or the evening service would run into the first morning service. No one went by the clock.

There were about ten teenage girls who lived out there and there were lots of blueberries. The girls would pick and send the berries to Arcadia. We got ten cents per quart. That's how we bought the cloth to make our dresses. We could sometimes find white sacks that cornmeal, flour, and grits came in at the store. I think they were ten cents each. We would make sheets, towels, and pillowcases from them after they had been laundered several times. We didn't ever think about ironing them. We never found out they were scratchy. It would have been too much trouble to iron them because we would have had to build a fire in the wood stove, put the iron on the stove, and wait until it got hot. Also, the house would get so hot, so everybody just wore wrinkles and all.

I did have one white dress that I made for special occasions. I was married in it. One time I went to Arcadia with my parents. Someone had opened a dry goods store near the grocery store. We girls heard about it. They kept a little of everything. Some girls bought a hat. Two of the girls asked me to buy them each a hat, so I did, paid for with berry money. I tried hats on and told the saleslady I would take them, and when the girls saw what I had picked out they didn't look happy. But they took the hats anyway. And they wore them to church.

We didn't have an inside toilet, so we used a big palmetto patch. There were two behind the church. Women used one, men the other. I never heard anything about any of the men forgetting which palmetto patch to use or any-

one peeping. No one broke the law at Hen Scratch. We didn't have any trouble about toilet paper—Sears Roebuck furnished it. One catalog would last for a three- or four-week meeting if no one had an upset stomach or if it didn't rain and get wet. If worse came to worst there were always plenty of palmetto fans—just strip off one of them and give it a twist and it would do until you could do better. But Sears was good about sending us toilet paper, and one other thing—they never sent a bill.

One service the preacher got up and said there was a young girl who was planning to run away and marry and if she did she would regret it. My friend Effie, who lived with her grandmother, had told me she was going to run away with a boy from Arcadia that night after the service. Immediately after church my papa saddled up the horses and loaded the wagon and left because he thought it was me. I didn't even have a boyfriend.

My brother married one of the Crews girls. The Crewses had eleven children. One child was struck by lightning, another turned a lamp over and burned to death, still another died mysteriously after swelling up around the neck. One son, born deformed, died at age seven, only able to make a terrible noise, and another son died of lockjaw.

Marilea Summerall, whom Rob married after his first wife, Cassie, died.

The old gathering place for camp meeting two weeks every July and Christmas was at Hen Scratch. The nearest town was Arcadia, thirty miles across the prairie. Each year it grew bigger, and finally, after the old-timers that lived around Hen Scratch died, the camp meeting moved to Ortona.

I had married a young girl, Cassie from Bonita Springs, that had come to Chokoloskee with a group of missionaries to hold a revival. She had been married to a man who had not treated her well.[4] In July of 1916 we were married at Bonita Springs at her home by David Leon McCormick. We lived in tents until we moved into our oyster bar house.

Cassie gave birth to our son Vance in that home. Her mother came with a midwife from Bonita Springs several weeks before Vance arrived. I was content to fish and come home to Cassie and Vance. We went from house to house fellowshipping with friends. That was our church.

First Ortona Church.

FIRST ORTONA CHURCH

Hen Scratch, gathering place for a two-week camp meeting held every July and Christmas.

In the summer of 1917, George and Lanora, Ed and Minnie Townsend, and Cassie and I were to meet Babe and Sally Whidden in Moore Haven for a meeting. We left Naples by boat, came up the Caloosahatchee River, and anchored our boat to the LaBelle bridge. We got somebody to drive us to Moore Haven, where Babe and Sally were waiting with the tent. We spent the night on a houseboat at Moore Haven and the next morning took a steamer to Pahokee. Land was secured and orange crates were to be our seats, but we were unable to find trees to cut for the tent poles. We decided to go on to Okeechobee City. It took us all night on the steamer from Pahokee to Okeechobee City. The pipes were rusted out and unable to gather up much steam. We stayed in the tent there, and Babe and Sally stayed with George Mansfield, who had a big house. Later, the mosquitoes got so bad we moved to George Mansfield's house too. I was tired of lying around during the day, and Babe said there was a big crop of new potatoes to be harvested and help was hard to get so he asked me if I wanted to dig potatoes. I was glad for the job. They had a pitcher pump in the field and the weather was hot and I drank plenty of water. I'm almost sure that is where

I picked up the typhoid germ that almost killed me about two weeks later. The owners were surprised at how many boxes of potatoes I filled during that day.

From Okeechobee we left for Hen Scratch for the Fourth of July meeting. The Holy Ghost had revealed to Sally that someone would die during the meeting. Everyone was scared. When I got there, I came down with the worst headache and fever I'd ever had. Old Aunt Missouri Williams would gently massage my brow and pray in the Spirit. The meeting ended and Cassie, Babe and Sally, and I moved into the old cabbage-thatched tabernacle because I was too sick to be moved. When we first got there we stayed in an old schoolhouse right beside the tabernacle.

I lost track of days and time because of a high fever. Everyone expected me to die except Brother Dave Durrance. He'd come faithfully every day and pray for me. We all had moss mattresses, but Brother Durrance had brought a real mattress and put it on the floor for me to sleep on. Cassie was sleeping near Babe and Sally, and I heard a sound and Sally told Babe Cassie was dying. That's all I remember. Cassie died in Babe's arms.[5] Gone so quick. That day she was well and she and Sally had a Jericho March around the tabernacle.[6] I remember nothing more except one

Cassie, shortly before her death; Rob; and their one-year-old son, Vance.

thing: Babe came and told me they were going to take Cassie away—did I want to see her. Ed Townsend lifted me in his arms and carried me to see her. She was lying on some board, and that's all I remember.

For ten days they said I ate nothing, but Uncle Dave was faithful to come pray every day. They sent word to my folks that I was dying. Papa and Mama came. Mama had to return home, but Papa stayed with me and never left. My mind will always remember Aunt Missouri's healing hands as she'd rub my back and soothe my brow. Her lips would always be quivering in prayer. Every single day Uncle Dave would repeat over and over that I was going to be all right. I slowly improved, but was unable to work. Ed Townsend would lift me to the toilet. The day came when they felt I could go home. Marion Crews had an open-body truck. He filled it with moss, and Ed lifted me in the back of that truck and they drove to Palmdale to the train station. We just made it before the train left. Ed carried me to the train and put me in a seat beside Papa. We went to Fort Myers, where my brother Claude met the train and carried me to his house. Claude wanted to know what I could eat and I told him I thought I could eat a mullet. He called his doctor and he said I could have a small piece of boiled mullet, but no fried. On fishhouse scales I weighed eighty pounds and began to gain about a pound a day for a month.

Cassie's parents came to Hen Scratch and took Vance.[7]

"THEY WERE A BIG HAPPY FAMILY AND SEEMED DIFFERENT THAN THE PEOPLE AROUND HEN SCRATCH." MARILEA

The next camp meeting we got ready and loaded up and headed back. We got there a day or two before it started. Papa wanted to be sure he didn't miss anything. Someone told us there were some people from a place called Everglade that were coming. It was down on the west coast. The night the meeting started there came six couples besides the children. They brought a big tent, stretched it up, and all made beds under it. They had a little stove with two burners—they called it a camp cookstove. It burned gasoline. We had never seen one before. How they ever got enough cooked for that many people I'll never know. They were a big happy family and seemed different than the people around Hen Scratch.

People were coming in from everywhere, and everyone camped around the tabernacle and cooked on an open fire, except the people from Everglade. I remember I would have to make bread and bake it in what was called a dutch oven. You put fire on top and under it, and mine always came out burned on the bottom and white on top. But Papa didn't have to tell me but one time he didn't want any more burned bread. I learned just to put a little fire under the oven. I told myself I'd never bake bread again in an oven outside when I get married, and I have kept my word.

There were two young couples from Everglade. One couple had a little boy. The father of the little boy was not feeling well, but he came to all the services and didn't act like he was sick. I didn't notice him anyway. The meeting lasted three weeks and we had to go back to Arcadia.

This man got so sick he couldn't go home. They got a bed and put it up in the tabernacle for him. One of the ministers and his wife

Old Marco Pass in 1915.

stayed in there at night to help care for him. One night this lady heard the sick man's wife making a noise. Her name was Cassie. Her husband was so sick that they didn't tell him she had died until the next day. He didn't know she was dead until weeks later. He didn't know anything.

His brother went back to Everglade, got his parents and they all came up and stayed until he got able to go home. Two men picked him up, put him on a mattress in back of a truck, and took him to Arcadia. Three more men put him on the train. His father had gone back home to Everglade, got his boat, and was waiting for him at Punta Gorda. His mother-in-law came to Hen Scratch and got his little boy. The sick man was down to eighty pounds. His mother took care of him and he began to gain and soon was well.

"PAPA ASKED ME WHAT THIS FELLOW WAS WRITING TO ME ABOUT, AND I SAID FISHING." MARILEA

We had gone back to Arcadia so Papa could work until the next camp meeting. It would be at Christmastime. One day some people stopped by our place and told us the lady that had the little boy died. She wasn't sick, she just died in the night, and they didn't think her husband was going to live. I thought, I am going to get that young man. I was engaged to a boy in the service, or so I thought, but I had heard a bird in the hand is worth two in the bush.

Soon it was Christmastime and camp meeting time at Venus. When we got to Venus he was there. The tabernacle was better than the one at Hen Scratch. It had a floor in it and a good top—also a board to lean on when you sat down. There weren't but two houses that you could see from the church, but all around the church was filled with camps. There was a little place where you could buy a few things if you had any money. We didn't ever seem to have much, but no one worried and everybody was happy. I still had to make bread, but believe it or not, just before I married I perfected baking bread in a dutch oven with Papa's help.

The first night I was supposed to lead the singing. I was looking through the song book and this young man came and sat by me. I think he spoke to me, and I guess I said hello, but that was all. After we married he told me the preacher told him to go sit by me. When we started to our camp after church that night we had my sister and a little girl living with us. She was asleep. I picked her up and he came over and told me to let him carry her, and we walked to our camp with Papa and Mama. Nothing else happened. He had to leave to go somewhere between Venus and LaBelle. His car had broken down and he had to get parts and fix it. I thought, Well, he could have said goodbye. The meeting closed and we went back to Arcadia.

A few days later after we got home and went to get the mail I had a letter from him. Nothing new, only he had started fishing and was doing good. At the end of the letter he asked me if I would write off a song that I had sung at the camp meeting. So I did. He wrote and thanked me and told me all about the fishing again. I thought, I am still going to get you, but you don't know it yet because I am going to make you think I am hard to get, so I will go slow. I'd wait longer to answer his letter, but I was sure to answer it. I wanted him to say something else besides fishing, but he never did. I would sometimes get two letters from him when we went for the mail. We didn't go but about once a week. Papa asked me one day what that fellow was writing to me about, and I said fishing.[8]

Vincent Braman and Chester "Junk" Pettit on their way to Punta Rassa with the mail (1916–17).

ABOUT 1916 OR 17

THE STORY OF VENSON BRAMAN AND JUNK PETTIT.
ON THIR WAY. TO PUNTARASSA WITH THE MAIL. THEY HAD
QUITE A LOAD OF TOMATOES, IN CRATES AND WAS TOP HEAVY
AND IT WAS A LITTLE ROUGH WITH A STRONG NORTH EAST WIND
AS THEY WERE PASSING THE BELL BUOY. A FEW HOURS BEFOR
DAY LIGHT THE BOAT TURNED OVER. VENSON DESIDED HE WOULD
SWIM TOWARD SHORE TWO OR THREE MILES. JUNN WAS SCARED OF
SHARKS AND STAID WITH THE BOAT. VENSON SAID THERE WAS LOTS
OF FOSFORSE IN THE WATER AND HE SAW LOTS OF SHARKS. AND
WHEN HE GOT TO WHERE HE COULD TOTCH BOTTOM ON CARBOS BAR
THERE WAS LOTS OF SHARKS. AND WHEN NEAR ONE HE WOULD STOMP
THE BOTTOM AND THEY WOULD LEAVE. BUT ALL THE TIME HE
COULD FEEL ONE BITING HIS FOOT OFF. BUT HE FINALY MADE IT TO
SHORE SAFE AND SOUND. ALL MAIL DOWN THE COAST AT THIS TIME
WAS CARRIED BY BOAT. FROM FORTMERS. BY PUNTARASSA. NAPLES, MARCO
AND EVERGLADE AND CHOKOLOSKEE. (CAPT. JOHN WEEKS WAS THE)
SOME TIMES THE MAIL WAS A LITTLE LATE) (MAIL CARRIER FOR MANY YEARS)
AS UNCLE JOHN HAD TO MUCH. TO DRINK
MON WED & FRIDAY THE MAIL BOAT FROM MARO TO EVERGLADE & CHOKOLOSKEE

FIVE 🐟 HARD TIMES

I have been asked many times what I knew about E. J. Watson. I did know him and a lot of things that he did. But somehow I never did like to write what I knew or talk about it.

Mr. Watson was a friend to my dad, as they both had little schooners and met at Key West and other places to sell the stuff that was produced in that part of the country, especially syrup. Mr. Watson's schooner was called *Gladiator*. He ate many meals at our table. He'd come to Everglade once a week to pick up groceries and his mail. He most always arrived about noon and would eat dinner at our house. Sometimes we'd see him at Key West, and he always came aboard our boat and visited. He didn't have a wife when I first knew him. I knew him as a friendly, jolly man. I didn't know what was down inside of him at that time. I remember him very well. He had a broad mustache—sometimes side whiskers.

I knew his boys very well, especially Lucius. We called him [Lucius] Major or Colonel because he was in the First World War. My Aunt Nannie practically raised and schooled Ed [Jr.] and Lucius because they had no mother. Lucius was like a brother to me.

Lucius married the widow Williams at Henderson Creek. Ed lived in Fort Myers and was a more educated businessman and into real estate. The Colonel was like me. He liked to guide and fish—that's why we had so much in common.

Well, all that dirty life and history of Watson [Sr.] didn't interest me. I knew the day he was killed and where and who did it. My brother George was in just yards of where it happened. He heard the gun and in minutes he saw the whole thing. They put Watson in a canvas bag and buried him out on Rabbit Key. W. G. Langford, his relation, had W. D. Collier move the body to Fort Myers to be reburied in the cemetery.

Mr. Watson had a big cane farm at Chatham River—about forty acres. He made lots of syrup. There was a bad happening one day—a Negro was grinding cane and his croker-sack [coarse cloth] apron got caught in the mill, and in trying to free it his hands were pulled in the mill and it ground his arm up to his shoulder and he right quick bled to death. For a long time after that Watson had trouble selling his syrup.

I was sixteen when the Watson killing happened. A strong northeast wind had been blowing for days, which was not uncommon, especially in October during hurricane season. Early on that mid-October morning Ed Watson came and asked my papa to take him to Marco. He said he was tired of all the nonsense about being accused of murder.

Watson had come to Half-Way Creek around 1892 and was supposed to have killed Belle Starr out in Arkansas.[1] He worked awhile around Everglade, then bought the Chatham River cane farm, raising sugarcane and vegetables for the New York–Key West market. Mrs. Hannah Smith, who had been engaged in hunting alligators with a Mrs. McLane from an oxcart, was killed about 1910 at Chatham Bend. Mrs. Smith and a man named Waller, who also lived at Chatham Bend, were murdered and their bodies weighted and thrown overboard. Mr. Watson was at Chokoloskee Island at the time they were killed. An escaped convict by the name of Lesley Cox from Madison County was also staying at Watson's place, as was another convict named Dutchy Melvin from Key West. The convicts were suspected, but no trial was held. Cox later killed Dutchy and escaped.

After the murders had been committed, Cox told a Negro who was working at Chatham Bend that he had been hired to kill Mrs. Smith and Mr. Waller and then kill the Negro. However, Cox decided not to kill the Negro and instead told him to leave the island. The Negro left and told several people, including Claude, my brother, what Cox had said to him.[2] As Cox, Melvin, and the Negro were employed by Watson, the people in the area suspected Watson of having the crime committed. When Watson heard that the Negro had left this vicinity and was on his way to Fort Myers to report the incident to the sheriff, he had my dad take him to Marco in an attempt to reach Fort Myers before the Negro. When my dad refused to take him further, Mr. Watson hired passage to Naples that night. On leaving Naples he was caught in the hurricane of October 18, 1910, at Bonita Springs. Upon finally reaching Fort Myers, Watson learned that the Negro had arrived ahead of him and that the sheriff had already departed for Chatham Bend to arrest Cox. Watson hired a speedboat at Fort Myers and overtook the sheriff at Marco, where he failed in an attempt to have the sheriff appoint him a deputy to get Cox. Returning from Marco, Watson went on to Chokoloskee, and a number of people,

including Charley Johnson, three transient fishermen, and Dan House, met his boat at Ted Smallwood's landing. Watson stated that he had killed Cox, and when the people asked for proof, Watson produced a gun and a hat, which he said belonged to Cox. At that point, Mr. House stated that he wanted more proof than that that Cox was dead and told Watson to give up his gun and go with them to look

at Cox's body. Watson grabbed his gun saying, "I'll give you my gun," and he pulled the trigger. Watson's gun failed to fire, as did House's gun when he attempted to shoot in self-defense.[3] Others in the group then opened fire on Watson and shot him to death.[4] It was learned that Mrs. Watson was at the Smallwood house at the time of the shooting.

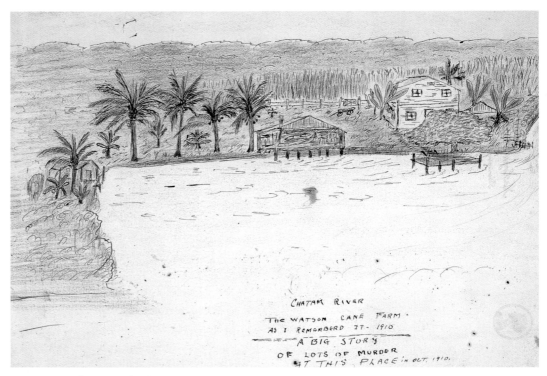

Watson cane farm on Chatham River.

We called him Mr. Watson or E. J. He was a two-hundred-pound, red-headed man, with a quick and sometimes quirky temper. Watson was reported to have been a hand of Belle Starr's and to have shot her in the back from ambush one night in Oklahoma. He was also said to have killed three men in Georgia and one in mid-Florida. He also cut the throat of Adolphus Santini in Key West, but Santini recovered. A deputy sheriff sent from Key West to bring Watson back was tricked out of his weapon by Watson and then put to work in the cane fields—under the gun—before being sent back, unharmed, but very respectful.

You know Knight and Wall Hardware in Tampa—Watson would sell his syrup there. Stayed sober, right sober, while working his crop, but at the end of the year he'd go to Tampa and get half drunk. He'd go in Knight and Wall's, and he'd have that pistol, that .45, and he'd see the clerk and say, "You look like a fellow that can dance," and pull out that gun and say, "Let's see you get right out there and dance," and the clerk had no choice. Watson would get three of them out to dance, and when old man Knight came down, Watson would turn them loose and say, "I didn't mean any harm, I was just having a little fun."

Back at Watson's cane plantation at Chatham Bend, inland from the Ten Thousand Islands, his ideas were serious business, no fun at all. With his reported past history of violence, his neighbors were worried, especially at reports of workers turning up missing.

He had a lot of hired help and he paid them for the year at the sale of his crop. He'd put a little money on the table and say, "Take this or this," pointing to his .45 caliber pistol.

I went up Chatham River mullet fishing the day before the blowup, and old man E. J. waved at me and a young man I had with me. The next morning Frank Cannon and his son, Jack, found the floating body of a woman who had worked for Watson. She was Hannah "Big Six" Smith, and she had two hundred pounds of pig iron on her, but she floated. The forty men on the clam barge at Clam Key called for action in the case. They suspected that Watson had hired Cox, an escaped convict, to do the killing. Both Cox and Dutch, and Waller, another Watson hand, couldn't be found.

Watson came to Chokoloskee just as the 1910 Hurricane was brewing and hired my dad to take him to Fort Myers to seek the aid of Sheriff Frank Tippins to arrest the murderers. They got only to Marco before they had to seek shelter. Tippins came south about that time, checking rumors of the killing, and missed them.

After the storm, Watson returned to Chokoloskee as a mob started gathering with guns. Watson hired Jim Demere to take him back to Chatham Bend, seeking Cox. They returned after three or four days, and Watson told the crowd of fifteen or twenty that he had Cox—that he had shot him out of his boat and he sank.

Harry Brown, leader of the group, said, "A body rises in twenty-four hours and we're going down and wait for him to rise; we want you to go with us." Charley Johnson told Watson to give up his gun. Watson declared, "I'll do no such damn thing," and he reached for his pistol, and then Henry [Short] shot him with a .25–.35 between the eyes, and then all the others shot, too. Uncle George ordered the men to go bury him.

The six feet of storm water over Everglade had spoiled all the water in the cisterns, and all the power boats were drowned out. So another fellow and I had rowed over to Chokoloskee to see if there was water there. When we were about three hundred yards away we heard guns go off, about twenty bullets, and a gang was passing the bottle around—they had been drinking pretty heavy—and we saw Watson lying there.

The crowd got the boats and tied a rope around his neck and rolled him into the boat like an alligator and took him to Rabbit Key where they dug a deep hole with no casket. They left him there with a flag on a pole over him. Watson had married a prominent Fort Myers girl, and the family paid Captain Collier to recover the body and return it there, where he's buried under a little old stone with no date or nothing, just "Watson," but everybody knows who it is.

Sid Linsy, arrested for helping kill Smith, Waller and Dutchy, was released by a Key West grand jury, and given money and a train ticket to his home in Georgia. Cox was never heard from again.

Despite Watson's notorious reputation, he was well liked by the Storters. His cane syrup, Island Pride, was some of the best around. He always paid his bills on time. Lieutenant Hugh Willoughby, a writer and explorer from Miami, hired Watson to guide him when he was collecting material for his book *Across the Everglades* (1898) and said Watson was the perfect host when he stayed at Chatham Bend.

"IN JUST MINUTES THE RIVER WAS OVER THE BANKS AND SOON IN OUR HOUSE." ROB

The first hurricane I remember was in 1910. Sixteen years had passed and we had not had a hurricane in Everglade. The wind blew for several days from the northeast and we had lots of hard rains. The people on Chokoloskee said looking up the bay it looked like a big mudflat.

Cranes and herons were feeding in the middle of the bay. The wind was blowing so hard out of the northeast that the tide couldn't rise. The Everglade River was sure low, too. But when the hurricane wind changed to the southwest the water came in like a tidal wave. In just minutes the river was over the banks and soon in our house. We were without water for a while. We borrowed enough to drink from Chokoloskee until we could get water from the head of the river. It didn't rain a drop for about three months after the storm. We had to haul our water until the next rainy season. We went to Gomez Island and got some water from the surface wells there. My dad and my oldest brother had gone to Marco the day before the hurricane broke to carry Mr. Watson after the last killing at Chatham Bend.

That night was a bright moonlight night—bright as day—and at ten o'clock my dad had still not returned. We worried. Uncle George came to my mama and told us we had to leave immediately and take refuge in an old cane scow with his family. He warned that the wind would change direction soon and would bring the water up on us like a tidal wave. We got to the scow just in time. The river in front of our house was almost dry, but in a matter of minutes the water came rushing furiously over the bank. My uncle's family, three Indians, two Negroes, another family, and my mama and us kids shared the shelter. We were worried about our papa, but there was nothing anyone could do. Sometime after midnight the rains stopped, but the wind carried the scow across a barbed-wire fence and lodged it between the mangroves. We watched as our little schoolhouse floated by us like a box. It lodged in the mangroves and tore to pieces. We saw cows, barrels, and boxes floating by. We spent the night in the scow. Around daybreak the wind subsided and we got down in five feet of water and pried the scow loose with poles and floated over to our house, four hundred yards away. The house was built a good three feet above the ground, and we saw mud eight to ten inches deep inside the house and the water mark on the walls reached a good eleven feet. Upstairs had not been heavily damaged, so we lived up there until the downstairs could be restored. We had an awful time getting the mud out of the house. My uncle found a cow in his house. We later learned that the hurricane had claimed one victim: a baby was swept from its mother's arms down at Pavilion Key.[5]

My dad and brother Claude, who had taken refuge with the Martins and their five small children, had spent the night in a small concrete cistern that had just been cleaned out.

My dad's boat, a thirty-eight-foot launch, had broken loose and lodged in the mangroves and sunk. He borrowed a row skiff and rowed eight miles to Everglade.[6] Many thousands of mullet died. The water was thick with mud. All of our chickens drowned. All of our fresh water was destroyed with the overflow of saltwater. For several days we went wanting for water. Some of the boys rigged up a moonshine still and made a little fresh water to drink until we could get some elsewhere. When the water was all gone we took a cane

scow up the river where the water was fresh and bored some big holes in the bottom and let it fill up and towed it down to our home and rigged a long gutter to our cistern. And with a pitcher pump, with lots of hard work pumping, we put enough water back in the cistern to last us until the rains came and put more water in. This was the most tiresome thing I ever did. The mangroves all around the bay were torn up and destroyed. It was an awful sight for a long time.

The schoolhouse, drifting by during the eye of the 1910 hurricane.

On my dad's way to Everglade after the 1910 hurricane, he and my brother were passing a little place called Fool's Point, where two old men had been camping. They heard someone talking. They looked up in a big old torn-up tree and saw two old men that had weathered the storm. Their camp, their boat, and everything they had was gone, but their lives had been spared. Claude and Papa took them in and brought them to Everglade. Mr. Jordan and Mr. Mount were visitors to the area and had set up temporary housekeeping on Fool's Point to do some experimenting with a pound net. When my dad passed and found them hanging desperately to a tree, their cabin and supplies were nowhere in sight. Mr. Mount, before leaving the island, chipped off a piece of the tree declaring that was all he ever wanted of Florida.

Mr. Mount returned to the north, and Mr. Jordan moved to Wood Key. He lived up a little creek where Gene Hamilton had a little farm with some rich hammock land. At times deer would come in there to feed. So Mr. Jordan set a shotgun for the deer. It was when coon hunters would hunt coon at night with a headlight. A hunter was hunting near there this night and shot a coon. Jordan heard the gun shoot and thought it was his gun and

rushed out to see if he had killed a deer and hit the wire and fired off his gun and shot his legs off and he bled to death. This was always a warning to me never to set a gun, although I have been tempted. I have a better idea: Set a monofilament string where the deer will run into it, and have it on a fishing reel and

the clicker on. Then get up and kill the deer. When the deer hits the line it will sound like you have got a big fish. You can be several hundreds of yards away in your bed, maybe asleep, but you will know it when that reel starts singing.

A TERIABLE SIGHT

WHAT WAS LEFT OF FOOLS POINT THE MORNING & AFTER THE 1910 HURICANE.

TWO OLD FISHERMAN HAD A LITTLE HOUSE IN THIS POINT, THERE IS A LITTLE SHELL MOUND HERE WHEN THE WATER IS DOWN BUT THE NIGHT OF THE TERRIABLE 1910 HURICANE THEIR HOUSE WASHED AWAY AND SO DID THIR BOAT AND EVERY THING THEYHAD. THEY SPENT THE NIGHT IN A MANGROVE TREE, THE NEXT MORNING BEFOR THE WATER HAD GONE DOWN, MY DAD AND MY OLDEST BROTHER HAPNED TO BE ON THEIR WAY HOME IN A BORROWD ROW BOAT, AFTER SPENDING THE NIGHT AT FAKACHEE WHERE THEY LOST THEIR LAUNCH. AS THEY WERE PASSING THIS PLACE, THEY HEARD SOME BODY TALKING UP IN A TREE. IT WAS MR. JORDAN & MR. MOUNT. MR. MOUNT WAS FROM THE NORTH SOMEWHER HE SAID ALL HE WANTED OF SOUTH FLORIDA WAS A BIG CHIP OUT OF THAT MANGROVE TREE TO TAKE BACK NORTH WITH HIM.
A BIG STORY OF MR. JORDAN AND HIS DEATH, YEARS LATER—

Two men in a mangrove tree after the 1910 hurricane.

"HE PRAYED FOR GOD TO HAVE MERCY ON HIM THAT DAY WHEN THE BOAT TURNED OVER AND DROWNED HIS THREE BOYS." ROB

A sad story. The *Speedwell* bottomed up in the Gulf of Mexico. This is the last schooner that Captain W. D. Collier made at Marco. He took special pride in it. He built several other schooners before this one. He was making his first trip to Key West with the *Speedwell*. He had three of his little boys and four more passengers on board. The boat was light with no ballast in the bilge yet, and the boat was in full sail and was top heavy and turned bottom up and drowned all on board except him and Cat Green, his mate, one deckhand, and one passenger. He was trying to make a fast record trip. Captain Collier claimed to be an infidel, but Cat Green said he prayed for God to have mercy on him that day when the boat turned over and drowned his three boys. But as far as I knew him he never changed.[7]

"I NOTICED A HUGE DIAMOND RING ON HIS SWOLLEN FINGER, BUT NONE OF US REMOVED IT." ROB

Cassie and I were living at Chokoloskee around 1915 in order to be nearer to the best fishing grounds. My brother George and his wife, Lanora, and Cassie and I were living in a big palmetto shack—Cassie and I on the west end and George and Lanora in the other end. One afternoon Cassie and I went bird hunting and did not get back until after dark. We were to meet George at a certain place after dark to start mullet fishing. George never arrived. We waited quite a while and he didn't show up. I knew something had happened or he would have been there. We thought some of the family had taken ill in Everglade. So we went on in to our camp, and George and Lanora were gone. Their table was all set with the food and nothing was touched. We didn't know the trouble until the next morning.

Mack Johnson, Alice, and their small boy were living just a few steps from us. Mack came over early and woke us up. "You must be the bravest man in Chokoloskee," he said. "Did you know there is a dead man lying near your door?"

The shots had scared George and Lanora off. They jumped into a rowboat and rowed about five miles to Everglade and stayed at the schoolhouse that night. Why we didn't stumble over him when we came in I don't know.

This was just after the bank robbery at Homestead, where twenty-four thousand dollars and a lot of diamonds and jewelry had been taken. I recognized the man immediately as Leland Rice. Leland had gone to school with us at Everglade. We found a pistol in his shirt, and his pocket was stuffed with money. Mack and I went for help. We returned with four men and held an inquiry. We knew he was one of the men involved in the robbery. We counted five thousand dollars in his pocket. No one wanted to be responsible for the money, so we took it to the captain of a fishing boat that made weekly trips to Punta Gorda. The body lay on the ground most of the day. We wrapped the body in a piece of canvas and buried it there on the island. I noticed a huge diamond ring on his swollen finger, but none of us removed it. When someone asked if a few words should be said at the burial, one man replied that his sermon had already been preached.

We found out later that two men had seen Leland come to the island, and when he wouldn't give them some of the money they had killed him for the reward money. Leland had come to Chokoloskee to get some groceries, leaving the others [his brother Frank and one Hugh Alderman] behind. He went near Raleigh Wiggins's fish camp, and Raleigh met him and shook his hand. Raleigh and Fred Hurst then followed him and hid behind our house. When Leland came out with the groceries he shot him with a load of birdshot. We heard that Leland, who had a pistol with him, did not reach for it, that he just looked up and Raleigh shot him.

When Leland did not return to his camp, Frank and Hugh swam to Chokoloskee and asked where Leland was. He was told Leland had been buried that day. Frank asked where the money was that was on him and was told that the boat captain had it and he could claim it because no one wanted anything to do with it. Frank jumped in George's three-horsepower boat without even asking to borrow it. At that point Fred Hurst, who thought Frank was coming after him for his role in shooting his brother, shot him through the lungs with a pistol, and he fell across George's boat and into the water. His cap, with his six-thousand-dollar share in it, drifted off and sank. Later, it was found by some local boys, who refused to acknowledge it.

Frank asked who had shot him. Fred walked over and shook hands with Frank who then asked Fred to take him to a doctor, and Fred responded that he would do the best he could. Frank was carrying on in the bottom of the boat, bleeding, telling kids that had gathered there not to ever become desperadoes. They took Frank to Sheriff Tippins and then to a hospital in Punta Gorda. He was sent to Raiford Prison, where he developed tuberculosis and was pardoned. But he was helpless and bedridden, and died.

They left Hugh in a house there at Chokoloskee. He had his six thousand dollars wrapped up in a handkerchief and two to three hundred dollars of his own money. He hired brothers Alfonso and Joe Lopez to take him to the sheriff and he gave himself up. He was only seventeen years old.

Tucker, the fourth robber, drowned in a squall after getting on a drunk, and his body drifted ashore. A Seminole, Charley Tiger Tail, found his body, and the ears, eyes, and nose had all been eaten by the blue crabs. There must have been one hundred crabs on him when he was found. George wouldn't eat blue crabs after that. The same man who buried the first robber buried this man. A couple of men tried for days to find this man's money, using nets to drag, but if it was ever found no one heard of it. The money had come out of Tucker's pants right in front of Smallwood's store.

Alderman confessed and was taken to Fort Myers and sent to Raiford, but was pardoned when he went insane and was put in an asylum.

These were all young men. I knew all of them but Tucker. They had taken a dinghy and left the owner fifty dollars for it and rowed to Rabbit Key. Leland Rice's brother-in-law Marion Cason, who had married Rice's sister, found a map that he said was where money had been buried on Rabbit Key from the robbery. He and Preston Sawyer went and hunted for the money, but all they found was empty cans buried between cabbage trees. Somebody had beat them to it.

"THEY WOULD BRING THEM OVER TO NEAR CAPE ROMANO AND KNOCK THEM IN THE HEAD WITH A BALL PEEN HAMMER...THEN THROW THE BODY OVERBOARD." ROB

There was a time bootlegging whiskey was a common thing. Lots of money was to be made if you went into it in a big way. Lots of boats were lost and lots of people, too. I knew two young men that were on a trip to Bimini for a load and have never been heard of since. And I knew others that had narrow escapes of their lives. In the long run it didn't seem to pay off. Carlton Smith said he had just made a load for Dan House and was on the way home coming through Marco River near the Curlew Flats when his engine broke down.[8] It was in the middle of the night, and he was tied up to the mangroves on the side of the channel working on the motor and suddenly it caught fire and the boat burned up. He got out in the mangroves and weathered it out until daylight. The mosquitoes and sand flies came down on him so bad he just couldn't stand them. He got overboard and started swimming toward Marco, about three miles. He made it, but he said that was a lesson for him and he found a better way to earn a living.

Mickey Brown [Rob's cousin, Aunt Toggie's son] and another young man were returning from a trip to Bimini with a load of whiskey. Mickey was down in the cabin asleep, and fumes from the motor had knocked him out and his foot had dropped over on the hot manifold and cooked it so bad he almost lost his foot before it was all over. His partner on the outside that was steering the boat smelled something cooking, and he looked down in the engine room and it was Mickey's foot. He was completely knocked out from the motor fumes that he couldn't feel a thing. They were several hours from home. But when Mickey came to he almost went into fits from the pain and he was a long time getting well.

One night a big booze operation was taking place at Everglade. One car was loaded and was on its way out. Deputy Sheriff Hutto approached this car for arrest. Hutto was shot and killed and the booze went on its way. No one ever knew who did this murder. No evidence.[9] When government agents found a load of whiskey they would break all the bottles and sometimes set fire to it. Once they found a boat with a thousand cases and broke it all.

Making moonshine was another problem. But there was lots of it made and drunk. One night Bill Gandees got hold of shine that Dick Williams had produced and after a few drinks he went blind. And a few hours later he was dead. My cousin John Brown could make what they called "Good Double Run Shine."

One morning early I was fishing out near Gordon's Pass. I saw the Coast Guard going by with a boat in tow. It was Elmer Aldacosta with his first load of whiskey, and his last. The Coast Guard cutter was lying in the mouth of Shark River as Elmer was passing by in the moonlight. They soon had him in tow and on the way to Fort Myers. Elmer got out of this lightly, as it was his first offense. The judge gave him some good advice and it stayed with him. Elmer was just out of high school.

Not long after I got typhoid fever and Cassie died and I returned to Chokoloskee there was a bad happening of whiskey and murder. Horace Alderman and Ode Roberson were in the bootlegging whiskey business in a big way. But soon they got into a bigger money-making business—hauling aliens from Cuba to Florida. They would bring them over to near Cape Romano and knock them in the head with a ball peen hammer after collecting their money. Then they would throw the body overboard. Ode told my brother the whole story after he was converted. He said it was so awful bad a scene he hated to even think of it. Ode went blind.

Later Horace was on one of his trips and was overhauled by the Coast Guard cutter. He almost whipped the whole crew before they finally cornered him. This was an awful battle with the Coast Guard. He was finally jailed and convicted and paid for his crime with his life. He confessed the whole thing before he died.

My brother Wes as a teenager was involved with these men early on when they were hauling aliens and spent a short time in prison for it. It was an awful time for my parents.[10]

Hard Times   **75**

"PRO HI'S WERE MORE INTERESTED IN THE BIG BOOTLEGGERS THAN THE SMALL-TIME BUYERS." GEORGE STORTER

George Storter talked freely about his bootlegging years:

There was a near tragedy at Everglade when a bootlegging friend of mine invited me down to Port DuPont to watch the delivery of an order for fifty cases of whiskey. Port DuPont is across the Barron River. We hid the car and then we saw headlights coming and jumped under a little bridge, just barely room to crawl in. These Pro Hi's—prohibition agents—drove up right over us and said, "There's a boat coming; if we knew the signal we could ring him in." But they gave the wrong blink on the flashlight and the man in the boat gave it the gun. They jumped in the car and drove after him. They fired eight or ten bullets, but he lay down behind the engine block and they missed him. After they left there was nothing to do but go home. My friend had lost his glasses under the bridge and couldn't see, and I had to drive him home.

Later, my friend made a deal with me. I lived at Rock Creek and added a room at the back of the house with only an outside entrance with a strong padlock on it. My friend said to give him the key and he would make it worthwhile for me. He would bring several hundred cases and hide the boat in the island. Then he would bring them up the creek in a smaller boat and hide them in the room. He'd pay me one dollar or two dollars a month each for one hundred bags. That was big money, and people would ask how I was getting along—that I wasn't working. I'd tell them I'd saved up my money. All that was required was that I leave the sacks alone.

Pro Hi's were more interested in the big bootleggers than the small-time buyers, and they tried to stay out of local affairs. Agents were moved every two or three weeks so they wouldn't be influenced by bootleggers. At one point the authorities put more energy into catching alien smugglers than bootleggers.

I worked as a bartender for Al Joslin from Cincinnati at one party at the Beach Club. His yacht was anchored only one hundred feet offshore. There were no fancy drinks, just highballs, Scotch and soda, and such—all free. So while they were dancing in the big ballroom upstairs I would make fifty to seventy-five setups and have them all ready, and when they came down I'd shoot it to them. About midnight I ran out of liquor, but there was plenty on the boat, so I took the skiff and got enough for another couple of hours, four sacks (twenty-four quarts), and came back. The clubhouse was all lit up and I didn't see anybody on the lawn. The next day a bunch of Pro Hi's came down to the dock and said they had seen me come from the boat last night and if they hadn't gotten out of the way I would have stepped on them. They weren't after me or the party, just the bootleggers who brought it over. They said, "We don't bother tourists."

"WE HEARD HIS LAST WORDS WERE 'GOD BE WITH YOU TILL WE MEET AGAIN.'" ROB

The following account is taken from the Coast Guard Magazine. *Although Rob told the story many times, he did not write it down.*

The Horace Alderman bootlegging incident was one of the most notorious incidents in Coast Guard history. The Coast Guard was ordered to carry a Secret Service agent to Bimini, where counterfeit American money was being used to purchase liquor. About 35 miles out of Fort Lauderdale, the crew encountered a suspicious boat, which stopped only after firing shots across the bow. The Coast Guard found 160 cases of whiskey on board and arrested Alderman and Robert Weech. Alderman somehow got to his pistol before it was over, and the Coast Guardsmen and Secret

Service agent were killed. The cook's eye was shot out. Alderman ordered Weech to burn the patrol boat, but the gasoline failed to ignite. When Weech had trouble starting the engine, the remaining Coast Guardsmen attacked Alderman and beat him with a rusty ice pick and paint scraper, then subdued Weech. Alderman reportedly cried, prayed, and begged for the Guardsmen not to hurt him. Weech testified against Alderman in return for a year-and-a-day prison sentence for violating Prohibition laws, and after two years of legal maneuvering Alderman was sentenced to death by hanging. More than a million dollars was spent on Alderman's defense, reportedly by a rich Palm Beach woman who Alderman was bootlegging for. It was decided the hanging would take place at the Fort Lauderdale Coast Guard seaplane hangar because local and state officials refused to carry out the sentence. While in jail, it was reported that Alderman cried, cringed, and got religion. He pleaded self-defense and swore he believed they were highjackers who had threatened him. His last words, sung in a broken voice through a black hood, were "God be with you till we meet again." Alderman was the first and only man ever hanged by the U.S. Coast Guard.

Although Alderman was hanged for the Coast Guard murders, he was known to smuggle Chinese coolies [persons doing heavy labor for little pay] into the country from Cuba and Latin America. Friends and family would pay fare money to Alderman, and he would in turn deliver the Chinese to points in the remote Ten Thousand Islands. At some point it became easier to kill them and toss them overboard. Alderman was accused of alien smuggling and out on bond when the Coast Guard incident occurred. Prior to that he had spent time in prison for rum-running, having been caught after hiring an undercover customs agent as a cook for his boat, and for robbing and clubbing Josie Billie, a Seminole chief.

"THE WIND BLEW ALL THE WATER OUT OF THIS BAY AND WE WERE SITTING ON DRY MUD." ROB

I was fishing with my two boys, Lem and Vance, at Chokoloskee and camping in a very small houseboat that we had made for our fishing trips. The fish and birds started acting up like they do when a storm was coming, so I took the boys to Smallwood's store and he said I'd better take my rig and get going, that a hurricane was coming. We spent the night of the Labor Day Hurricane in 1935 in this little lighter—ten feet long, seven feet wide. It was in this little mud bay down Rabbit Key Pass near Chokoloskee. We cut three long mangrove poles and worked them down as deep as we could in a tripod form and tied them together at the top and tied our little houseboat to it. We had our launch and two fish skiffs to look after, too. I bailed a lot of rainwater out of them through the night. We were well protected by the mangrove trees, but the next morning the trees were all bare of leaves.

Vance and Lem, teenagers at the time, slept most of the night while I watched the barometer and kept the boats bailed out. It rained an awful lot and kept me busy to keep everything afloat. In the first part of the night the wind blew all the water out of this bay and we were sitting on dry mud, but when the wind changed it brought twelve or fifteen feet of water. I kept watching the barometer and I watched it all night and could tell when we were in the eye of the storm and when it had passed by. That was the lowest I ever saw a barometer.[11]

I told the boys that night that this was really a bad storm. It tore up the whole shoreline from Chokoloskee to Cape Sable. A fisherman that lived down in the Keys told me that he and his family and a few others—as many as he could get in the car—weathered it out and the water came up in the car. It was completely drowned out. They kept the headlights on and every minute someone would grab hold of the car on the outside and then get washed on and drowned. He said the next morning after the storm passed by you could see dead people hanging up on the trees with no clothes on and some floating. He said it was an awful bad, sad-looking sight.[12]

When we had radios we could always get warning enough to prepare our boats and nets in a safe place, then get on higher ground in a safe house with some neighbors. In 1926, Marilea, Lemuel, Marguerite, and new daughter Olivia, and I were living on a houseboat for the summer. We heard over the radio about the storm. Anticipating this, we tied the houseboat to a stake and took refuge on the east side of Chokoloskee island with the Charley Johnson family. Cap Daniels had brought his boat to where we had tied ours and swam ashore after securing his boat and almost drowned. He was not aware that a skiff and net had sunk near the place he had anchored his boat. The storm swept over the houseboat and jammed it between two trees near Claude's camp. The windows blew out and drenched everything inside. When we returned to Naples after the storm we discovered our house had lost its roof and Marilea's new organ was ruined. My dad had taken refuge in the Bayshore Hotel and watched the water rise to the three-foot mark. His oyster house and all his belongings had been washed away. My brother George had a real nice tent fixed up; Claude had a camphouse he had just made. The next morning after the hurricane they had nothing. Claude lost his boat and net. That broke us up fishing at Chokoloskee for that year. The 1926 hurricane brought a drought and there were a lot of brush fires that next summer.

Hurricanes are bad dreadful things. They seem to kill all fish that are in their path and ruin all the trees and fill the bays with debris making it almost impossible to net fish for awhile.

1935

Weathering the 1935 Labor Day Hurricane in a little mud bay at Rabbit Key Pass.

Fish camp at the southwest corner of Chokoloskee before the 1926 hurricane.

Fish camp at the southwest corner of Chokoloskee the morning after the October 1926 hurricane.

I HAVE PASSED BY THIS SCENE
MANY TIMES.

ONLY THE MEMORIES LEFT

RH

BIRDS COMING IN, FOR THE NIGHT
KNOWN AS THE Duck Rock

IN 1935 = ESTIMTED 40 THOUSAND BIRDS EVERY NIGHT.
ON THIS ISLAND.

Birds sleeping on Duck Rock.

SIX 🐟 HUNTING, FISHING, AND NATURE

I started serious fishing in 1909. Dad made my brother and me a thirteen-foot skiff and bought us a cotton net. We started with that and soon we got hold of another skiff and net and really went at it. First, we salted our fish to be sold in Key West and eaten in Cuba. In 1910, we started ice fishing.[1] It wasn't very long until we were catching as many fish as the old-time fisherman.

In my early days of fishing, Nelson Noble and I were fishing one night near the mouth of the Everglade River. The moon was shining about half moon. We saw a black object coming our way and heard a trout strike or making noise. Nelson said, "That's a school of trout, let's strike." We took them in and beat up good on the boat bottom to scare them into the nets. When we decided to take up and got hold of the ends of the nets they were so full of cats [catfish] we just couldn't clear them. We lost both nets.

In later years, George and I lost a whole net at Joe Grasses in 1925. In 1935, we lost another net so full of cats in Henderson Bay. Norman Santini and I almost lost a net at Johnson Bay in 1938. We cleared it, but it was really full of cats. I saw Gus Bishop come in one morning with a boatload of cats roped in the net. He cleared all day and part of the night to save his nylon net. He got his son to help him. In 1985, I almost lost another net. I cleared cats all day and finally roped in and came in and lost the rest of my net. Just so many cats I couldn't clear it.[2]

Harry McGill and I took a trip around Lake Okeechobee in a two-horsepower, one-cylinder, two-cycle, jump-spark motor. This was before the lake was drained and diked and farmed. The lake was full of catfish and alligators. Many catfish were seined and hauled to Fort Myers. There were no towns on the lake at that time, only sawgrass and alligators. There was one little camphouse by a lone cypress tree where Moore Haven is now, and a few old catfish camps on the lake.

In the early summer of 1914 Harry and I had a little skiff launch. We started from Everglade, went up the Gulf of Mexico to Punta Rassa, then up the Caloosahatchee River to Lake Okeechobee. We camped at the lone cypress tree on the bank of the river. There was a little corrugated tin house there—I suppose for campers passing by. We spent the night there. We met up with an old alligator hunter and invited him to go with us as he was acquainted with this part of the country. The next morning we started out around the west side of the lake. There were lots of old abandoned catfish camps and old seine spreads. Each day we would move a little more around the lake and hunt alligators at night. We killed 125 gators that paid our expenses and [made us] a little pocket change. This was a real adventure for us. We didn't see or pass another boat on the entire trip around the lake. It was all sawgrass for miles and miles around the lake, lots of hyacinth and fresh water and grass around the edge. It took us one month to make this trip back to Everglade. There were lots of ducks. This was still in the days of a stern-wheel steamer. We spent the night on one about halfway up the river. It was tied up to the bank. I don't know what for, but the men asked us to spend the night.

Harry married my sister Eva. He suffered with stomach pains for ten days and the doctor told him to get to a doctor in Fort Myers. He didn't, and within a week he died of a ruptured appendix. Just before he died I could hear him scream across the river as several men tried to hold him down.

One day I was coming up the beach between Caxambas and Marco. There was a black whale on the beach. I thought I would go close in and take a good look at him. When I got pretty close I thought I saw another live one, but it was a big shark about the same size. He smelled the blood of the whale. That's why he was there, but there was nothing he could do about it. But I gave him plenty of room. He was about as big as my boat, which was about twenty-two feet long. I sure didn't like the looks of him and didn't want to disturb him. My cousin Frank Brown told me that when he lived at Punta Rassa as a teenage he saw the water turn bloody as sharks attacked a school of mullet about a mile off shore.

When I was eight years old I begged Papa to let me dive off the *Bertie Lee* with my brothers and swim after the boat. Papa told me I couldn't swim fast enough yet, that I'd better sit on deck and watch. I knew there was no need to beg, so I settled myself to watch the other boys dive from the bow and swim to the stern and get into a small dinghy we were towing. This went on for a good half hour when I spotted something black behind the small rowboat. I told Papa that I believed we had hit some porpoises, that I saw something black following the dinghy. Papa took one look and yelled for the kids to get in the boat quick. It was a shark. Claude and George lost no time scrambling into the dinghy. We could now see the shark and it was about twenty feet long, about the length of the skiff.

Comer Island, about six miles out of Everglade, was named for the Governor of Alabama, who visited the island for many winters.[3] Walter "Chink" Bostick and I would guide the Governor and his family every winter for several weeks. Chink married my cousin Frankie. The Governor and Mrs. Comer would spend three or four weeks on this island every March. Sometimes he brought his daughter and her husband with them. We went fishing every

Rob, Jake (Governor Comer's cook), and Walter Bostick, who married one of Rob's cousins.

day, hot or cold, rain or shine, and caught lots of fish. What we didn't eat, Walter and I would sell. One of us would go in to the Chokoloskee fishhouse every night with 150 to 200 pounds or more of fish. We would get mail, water, groceries, or whatever we needed. We didn't get many groceries because the Governor brought a good supply with him. He brought his Negro cook, Jake, with him—a fine fellow and cook. He would bake blue-bill ducks and fish and potatoes in hot ashes in the ground.

The Governor would get up at daybreak and give a big war whoop. That meant for everybody to get up and get ready for breakfast and start fishing. He would go in bathing every morning at sunup, regardless of how cold it was. One time, after a freezing northwester, he only went out knee deep.

The Governor did this for several years until he died. After the Governor died, his son and daughter-in-law took over, but soon the daughter-in-law died and I heard the poor man tried to drink himself to death. Several times he tried jumping overboard while the boat was moving, but we rescued him each time. Finally, we heard he died of some kind of poisoning.

Pelican Key, later named Comer Island.

"THERE WOULD BE A COON IN THE TRAP . . . WITH HIS FOOT ALL SWELLED AND HE WOULD BE WIPING THE SAND FLIES OUT OF HIS EYES." ROB

Coon hides were worth lots of money to the hunters. There were lots of coons, especially in the Ten Thousand Islands. They would hunt them at night with headlights and shoot them, also trap them. Some of the old-timers looked forward to the time of year when they could start trapping. The first cool spell in the fall, the hunters would start getting ready for the winter's hunting and trapping season. It was interesting to me to see them getting ready, getting their boats and traps all in shape. This was quite a moneymaking thing for the hunters. Coon hides sold for fifteen to twenty-five cents and went up to fifty cents in

the 1930s. Otter pelts, another popular hide, went for eight to ten dollars. Coons liked the crabs that hid under oysters—that's how coon oysters got the name.

There was a fisherman named Morgan that had a camp right next to me and George. He was a hook-and-line trout fisherman. One season he decided he could make more coon hunting then he could trout fishing so he fitted out as the others did and tried it that season, but the next season I noticed he wasn't so enthused and wasn't getting ready as the other hunters were. I mentioned it to him one day and he said, "Rob, I tried it last year, but I don't want to do it anymore. I guess I am just too tenderhearted." He said he would go to the

traps every morning and there would be a coon in the trap that had been there all night with his foot all swelled and he would be wiping the sand flies out of his eyes with the other paw and looking up at him and saying good morning I am so glad to see you come to help me. Then he would get his hardwood club to hit him a few licks until he had busted his skull and then go to the next trap and do the same thing over and over. He decided he couldn't make a good trapper. He would rather fish and make less money. He just didn't have the heart to just keep taking a life for his hide—there were many other ways to make money for him than killing coons.

"THE POMPANO WERE SO THICK, A HUNDRED IN THE AIR AT A TIME." ROB

In 1915, we got five cents a pound for pompano. I have caught them for nine cents and hauled them to Fort Myers by boat with a five-horsepower motor. In 1978 we got $2.50 per pound, and $3.00 in 1979.

"One-Arm" Wheeler told me he came down from Fort Myers to Doctor's Pass pompano fishing. He had a greenhorn fisherman with him that had never fished for pompano. The tide had just turned out. When he got to the

Doctor's Pass bar the pompano began skipping and he told his partner to get out on the beach with the end of the net and hold on to it. He said he ran his net out and the pompano were so thick, a hundred in the air at a time. He looked back and the fellow had turned the net loose and was running down where he was, and all they got was what hit the net while he was running it out. He only got two hundred pompano. He said it made him so mad he felt like killing that feller. But he gave him a good cussing instead.

Many years ago when George and I had just moved to Naples we had never done any pompano fishing and there were lots of pompano but not much sale for them. They were compared to jacks. We had a good launch with a five-horsepower Palmer Motor. We made a few trips to Fort Myers. We got nine cents a pound for them. That was where we got acquainted with Mr. J. W. Riggs and fished for him many years. We bought a trammel cotton net from an old fisherman at Marco. If we had the kind of pompano nets that we have today we would have made some record catches. John Hogue from Wiggins Pass would catch the moon and tide just right and strike the Gordon's Pass bar at high slack tide and often catch five hundred pounds. He had a good net, but still not as good as we have today. Andrew Weeks told me once he was rowing a skiff with a silver mullet net on it, and as he was going in Little Marco at the mouth of the pass the pompano were so thick they skipped in his boat and out on the beach. He made a

land strike and pulled his net out on the beach and landed nine hundred pounds.

In 1978, a kingfish rig with a nylon kingfish net caught about nine to ten thousand pompano down off Cape Romano. It was a net with rings and rope and you could work it down like a purse seine. They forced them into it and got all of them. That was sure a lot of pompano for one outfit. They catch kingfish the same way. A plane spots them and directs them just how to make the catch. They killed the goose that laid the golden egg.

Pompano, a gulf fish found in deep water.

"THE BIG BLACK DRUMFISH MAKES SO MUCH NOISE YOU CAN FEEL THE VIBRATIONS." ROB

George and I caught a seventeen-foot devilish in 1917. For awhile it looked as if he would win, but when we got him fastened to the launch and pulled him backwards full speed we soon had him. He grounded in two feet of water and soon gave up.

There are a lot of fish that people are afraid of, and some for good reason. I wonder why some things are such a pest to the fisherman, like the spider crab. They are a real nuisance when fishing for pompano when you are in a hurry. They have sharp barbs all over the body and hard sharp toenails on the legs. Drums and redfish love them. Drums make a noise at night when you are over a school of them. The big black drumfish makes so much noise you can feel the vibrations and if you hear them you better get your net up in a hurry or you

are in trouble. Black drums can weigh up to ninety pounds. Drums sure like spider crabs. Drums root in the ground just like hogs root for worms and roots. Drums root for small-type clams and other things in the bottom, and they sure won't turn down a stray spider crab. How do I know these things? I have cut them open and seen what they were feeding

Sheepshead.

on. Pompano root, too, for cockle (look in his belly).

When you are fishing at night along the shoreline for snappers and sheepshead, if you listen close you will hear the singing toad if he is around. Probably you wouldn't know what it was. Once in a while you will catch one, but handle him carefully with pliers for he will sure bite you or you can get stuck with one of the sharp fins on his back and it is painful. There is another real poisonous fish. He is red-colored and has sharp spines and looks sort of like a little red grouper [yellowfish grouper]. Beware of getting pricked with a sea scorpion's sharp poisonous fin. There is more poison on one of his spines than any fish I know of. Very painful.

Hunting, Fishing, Nature   85

My brother George and I, Jack and Cap Daniels, and one hired help salted mullet at Dismal Key in December of 1925. It was closed season on mullet, but the law was not enforced and Sheriff Manard was on our side. We had Mr. McKinney get us two hundred sacks of coarse fish salt (160-pound bags). There were several other crews putting up salt mullet too. Fish were plentiful. One morning there were ten fishermen waiting for the same school of mullet to come through what was called Little Bay. About sunup (seven or eight o'clock) three schools began to come through into one big school. We put four nets around them and six in the center. I have never seen such a bunch of mullet, and what a sight it was when they found they were inside of a net fence. In a few minutes every net was as full as it could hold. We estimated ten thousand head. There were so many fish in this school, in a few minutes all our nets were sunk down with mullet and the water was as muddy as soup. It was really a sight to see when the fish found they were rounded up. It looked like a thousand mullet jumping in the air.

We worked all that day, all night, and all the next day on our share of the fish (about four thousand head). We only stopped long enough to eat, and then back on the job. One of the other crews brought over to our camp about three or four hundred head, as they were out of salt and we had to take care of them. When we got through with that batch of fish I have never been so tired in all my life. In ten days we put up one hundred barrels of salt mullet. We hired Riggs Fish Company to haul them to Key West for us. We got four and a half cents a pound for the mullet and ten cents for the roe. We just about lost all our fingernails. Poor old Cap, after the night and two days' work, he went home to get some more salt. When he got home he was fumbling on a dresser and got Annie's hair pins under every sore fingernail, and that broke him of ever salting mullet. I have helped land one hundred thousand pounds twice in my lifetime.

Seine hauls in December at Naples.

Thousands of birds, especially curlews, slept on Duck Rock Rookery for many many years. I saw curlews, herons, cranes, squawks, cormorants, and some mallards there. A warden told me that he estimated about forty thousand birds slept on this island every night. It was protected by wardens, but Hurricane Donna in 1960 tore it up and washed away the mangrove trees, and birds stopped using it and moved to other islands. I have passed by this rookery at sundown—it looked like a white sheet spread over it after they were all settled down for the night. You could see flocks a mile long coming in. There was another island called Rock Hole Island made up of several little islands. The Lopez brothers that were older than me told me at one time when they were small boys that this was just a lot of bare rocks and oyster bars. I suppose it consisted of 150 acres once and was bare just for a few small mangrove trees. Now it looks like one big island with tall trees and small creeks running through it. I have caught lots of mullet around this island in the creeks and holes. Every day between Marco and Goodland Point I could see about three hundred pink spoonbills feeding on a mud bank at low tide [from 1900 to 1915].[4]

Spoonbills feeding at low tide.

"HE SHOT THIS BIGGER ONE AND WOUNDED HIM, AND HE CAME STRAIGHT TO ME AND RAN BETWEEN MY LEGS." ROB

One night Nelson Noble and I were hunting alligators. He always did the shooting and I did the sculling. When he would spot one he would point his finger and guide me up to where he would shoot. This night I saw him point. He whispered real loud, "Boy, he's a big one." I sculled to him slowly and quietly. He shot and said, "I got a seven-footer this time!" He took him by the nose and lifted his head out of the water. He said he believed he couldn't get him in the boat by himself. But when we got him in the boat he was only six and a half feet long. Something or someone had cut his tail off just behind his hind legs. His head and body were big enough for a nine

or ten-footer. How disappointed we were, as we got a fancy price for seven-footers.

George and I have caught several big gators in our nets while fishing at night for mullet. We would tie their jaws together where they couldn't bite or fight and then you could do whatever you wanted to with them. We would bring them in and give them to some old-time gator hunter. He would skin him and sell his hide. Most always we would give it to Tant Jenkins if he was around. He was a good gator hunter.

George and I went on a deer hunt one evening in 1935 near Whiskey Creek.[5] We had a headlight with us.[6] It turned into a gator hunt, since we found a little lake that night just be-

fore dark and there were gator heads all over it. We decided we would wait until dark to kill them. George would shoot them and I would take them by the tail and drag them along until I got both hands full and carried them to shore. After we had killed most of them there was a bigger one that we hadn't killed yet. Finally, I was just behind George with my hands full. He shot this bigger one and wounded him, and he came straight to me and ran between my legs. I dropped those I was towing and was ready to get out of there. We had killed twenty-two. We went back the next day and skinned them. We thought this was a pretty good catch.

"I DID SOME EXTRA FAST SCULLING. WE WERE TWO FRIGHTENED BOYS." ROB

Nelson Noble came up to George and me as we were making a strike between Chokoloskee and Sand Fly Pass and asked us if we had anything to eat in our launch, which was anchored just a few hundred yards from us. He said he was so hungry, that he never could fish when he was hungry. We always had a box of canned food in the launch, which was hardly ever used. It was there for an emergency. I told him to go look in the grub box and he would find something—some pork and beans, corn beef, or something that would quench his hunger so he could go on fishing.

Well, he went to the launch and was gone quite some time. And after a while he came back and said he thought he would go home. "I never could fish when I was so full," he said.

Nelson and I did a lot of alligator hunting up Left Hand Turner River. We were going up this narrow creek in the afternoon about three o'clock. It had just rained ahead of us while we were going up this creek where the trees were grown over the creek. In a certain spot that we were passing we could smell a panther very strong. We went on up the creek and finished our alligator hunt, but had no luck

and were coming back out. And it was very dark in this creek. Nelson had the headlight in his hand instead of on his head. The gun was lying down in the bottom of the boat. I was sculling the boat. I said, "Nelson, this looks like where we smelt the panther." And just at that time the panther jumped out of an overhanging tree right by the boat. Now, this was an exciting moment. Nelson scrambled for the rifle, and as soon as he got it in his hand he did some fast shooting in the dark. And I did some extra fast sculling. We were two frightened boys.

John Brown, my cousin, was a genius. He could do anything and play any kind of stringed music. When he was just a boy he made a fiddle with a cigar box and could play it. He was a good carpenter and could build anything. We called him Panther Brown because he had some good experiences with panthers. The first panther he killed was over six feet long. He was a good hunter. I was with him when I saw him kill a big buck deer. He killed a bear up Lane River. He killed a panther near our home at Everglade. He killed lots of coons and gators. He was a good tanner with alligator hides. He made some real good belts and sold them. One time he heard some dogs barking across the river and thought they probably had a coon up in a tree. He took his rifle along and was a real good shot. When he got close to the tree he saw a big panther up in the fork of the tree. When the panther saw him he made a jump, and John put a bullet through him. Before he hit the ground he slashed one of the dogs across the back with one of his claws. John had to sew it up. But he killed the panther. Another time he set some coon traps just in back of his house in a hammock. One morning he was going to look at his traps. One had a panther in it, and just as he got near it the panther began to rage. He had been chewing on his foot, trying to get out. He pulled out and left his foot in the trap and got away.

John was good at making moonshine. They said he could make the very best shine, but he had a bad fault: he drank most of it. And it caused a cancer in his stomach that finally killed him. When we were boys together he always had a desire for a new adventure and wanted to run away from home. He almost persuaded me one time. Finally, he did run away and went down to the clam bar at Pavil-ion Key. He stayed away from home several months. When he came back he had learned lots of new things and habits that I didn't know. Once he said, "Rob, I have learned how to dip snuff, it's better than chewing tobacco." He also had the cigarette habit. One day he was going to give a lesson how to use snuff. We were out on the sugarhouse dock where we thought no one could see us. He said, "Now this is the best way. Take a mangrove leaf and fill it with snuff and pull out your bottom lip and pour it in." Well, I was taking my first lesson. I had the snuff all ready and my lip pulled out and was ready to pour it in. But something told me to turn my head and look behind me. There was Mama. I just dropped the leaf and snuff and waited till a later time to try it. Well, I did lots of things in my early days that I wouldn't do again if it was all to do over again.

Graham Whidden of Immokalee had a real big panther that killed several of his cattle. He'd been on the lookout for him a long time. One day he ran across him and he was prepared for him and killed him. He brought him in to Naples and gave him to Dr. Baum to be mounted. He is now in the museum there. My dad kept him in the ice room at his fish market till he was taken care of.

Jack Daniels was hunting out near Fakahat-chee Swamp for deer one evening just before sundown. He ran across two big panthers playing. He said he was almost too scared to shoot, but he started shooting and killed one. The other one got away. He said he decided to cut off a foot of the dead one to take back to camp for evidence or no one would believe him. His brother Fin said he never heard so much fast shooting. Jack said while he was cutting off the foot of the panther he could feel the other one jumping on his back. He was so shook he could hardly use his knife, but finally got one foot and took off.

At the same place Jack killed this panther, Bob Barkley and I went hunting once.[7] One evening we separated and he shouted that he had shot a bear that was up on the side of a dead cypress tree pulling off the bark for worms. The bear fell in the water, but he finally got away. It was getting dark. I told him we had better get out of there. I didn't want to be in there with a wounded bear. When we got out on the outside of the swamp I heard some crows hollering just a little ways from us. I told Bob if he hurried ahead of me to where the crows were he would find a deer. As soon as we had separated I heard a panther cry a couple of times and it got closer to me. He crossed between me and Bob and about that time I heard Bob shoot. He killed a deer, but it went into the swamp. It was about dark and he couldn't find the deer. I put on my head-light and saw a pair of eyes in the swamp. Thinking about the panther, I decided to take a shot at it with my Remington. I shot, heard a lot of commotion, but it stopped. I went into where it was right on the edge of the swamp and there was a big dead buck. All the time I was thinking about the panther. In a few minutes Bob came back. I said, "There he is if you can tote him. I can't."

The next day Bob wanted to go back and look for the bear. While looking, we found the deer that he had shot the night before. He was spoiled and the buzzards had eaten most of him, but there was a real nice rack of horns. Well, we got the horns, but Bob just couldn't give up looking for the bear. I told him I would go on out and wait for him. There was lots of water, almost knee deep. I went on out, leaving Bob in the swamp. I heard him shoot, followed by a pitiful holler. I thought probably a snake had bit him and he had shot it. Something had happened. When I got to him he was hanging over a dead cypress log. I didn't know whether he was dying or what.

He said he had shot himself. His gun and deer horns were in the water beside him. All I could do was pray. Bob was a heavy person, too heavy for me to tote. And it was a couple of miles to the Jeep. It would have taken several hours for me to leave him and get help in Naples. I said, "Bob, you have preached divine healing and miracles." I picked up his gun and deer horns and took him by the hand and said, "In the name of Jesus, get up and let's go." Soon Bob was walking slowly and singing in a language I had never heard. He kept walking faster until I could hardly stay up with him. He was shot with a pistol he was carrying in his belt. It got caught in a vine and shot him through the upper part of his thigh. He went to the doctor as soon as we got in and he said it was a clean hole—nothing to do. Well, it was a miracle to me.

"A FRIEND AND I CHASED A RATTLER SWIMMING IN ALLEN'S RIVER AND IT BEAT US TO SHORE." ROB

Diamondback rattlesnakes are not known as swimmers, but returning home one day after a fishing trip to Marco I saw one swimming across the channel. I stopped the boat and killed the rattler with my pole and held it up for the couple I was guiding.

My dad told me of seeing one swimming out of sight of land in the gulf between Cape Sable and Key West. He had stopped at Cape Sable lots of times to get some coconuts, and every time he went ashore he saw a rattlesnake and sometimes two. That was at a time when there had not been a hurricane in several years to wash them away. He wondered if the current had carried it out there or if he was lost trying to swim to another island. A friend and I chased a rattler swimming in Allen's River and it beat us to shore. We threw a guava can at it anyway. Rattlesnakes swim fast and can coil up and float like a life preserver.[8]

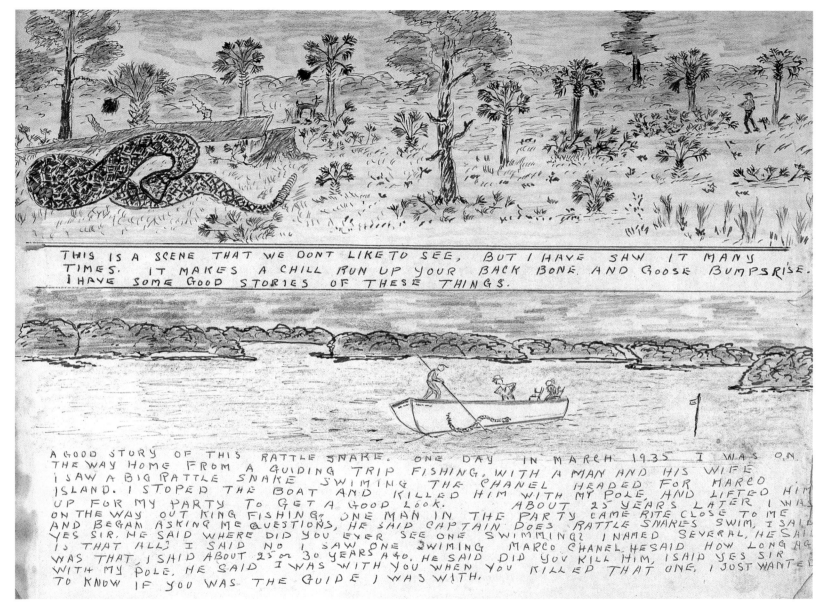

THIS IS A SCENE THAT WE DONT LIKE TO SEE, BUT I HAVE SAW IT MANY TIMES. IT MAKES A CHILL RUN UP YOUR BACK BONE AND GOOSE BUMPS RISE. I HAVE SOME GOOD STORIES OF THESE THINGS.

A GOOD STORY OF THIS RATTLE SNAKE. ONE DAY IN MARCH 1935 I WAS ON THE WAY HOME FROM A GUIDING TRIP FISHING, WITH A MAN AND HIS WIFE I SAW A BIG RATTLE SNAKE SWIMING THE CHANEL HEADED FOR MARCO ISLAND. I STOPED THE BOAT AND KILLED HIM WITH MY POLE, AND LIFTED HIM UP FOR MY PARTY TO GET A GOOD LOOK. ABOUT 25 YEARS LATER I WAS ON THE WAY OUT KING FISHING, ONE MAN IN THE PARTY CAME RITE CLOSE TO ME AND BEGAN ASKING ME QUESTIONS, HE SAID CAPTAIN DOES RATTLE SNAKES SWIM, I SAID YES SIR, HE SAID WHERE DID YOU EVER SEE ONE SWIMMING? I NAMED SEVERAL, HE SAID IS THAT ALL? I SAID NO I SAW ONE SWIMING MARCO CHANEL HE SAID HOW LONG AGO WAS THAT, I SAID ABOUT 25 or 30 YEARS AGO. HE SAID DID YOU KILL HIM, I SAID YES SIR WITH MY POLE. HE SAID I WAS WITH YOU WHEN YOU KILLED THAT ONE, I JUST WANTED TO KNOW IF YOU WAS THE GUIDE I WAS WITH.

Rattlesnake swimming the channel.

"IT LOOKED LIKE THE MOSQUITOES WOULD KILL US BEFORE WE COULD GET UP OUR NETS." ROB

I have seen mosquitoes as bad as they could possibly be, so bad it was impossible to fish at night. One time my brothers George and Hub were fishing together, and Lem and I were fishing together. George and Hub went up Chatham Bend River to Chevelier Bay mullet fishing and the mosquitoes ran them out. They told Lem and me that there were lots of mullet in the bay, but it was just impossible to fish in there for the mosquitoes. I told them I had never seen mosquitoes so bad that I couldn't stand them if I had some mosquito dope and if there were lots of fish. So Lem and I went up there. We made one strike and it looked like the mosquitoes would kill us before we could get up our nets. We just had to give up and get out of there.

Mosquitoes made fishing almost impossible sometimes. There was no mosquito dope in the early days, so the men would fill a bucket half full of dirt and black mangrove wood, then burn it. Sometimes they would put a rag in it and pat it down until it got to smoking good. These were called smudge pots. It would drive the "swamp angels" away.

"THE PINK SPOONBILLS FEED UP TO FIVE AND SIX FEET OF THE BOAT AND FOLLOW THE TIDE OUT." ROB

Dry-stop mullet fishing is done at spring tide.[9] At high tide the fish go into the grass and even back into the mangrove swamp. I put the nets out and put stakes on them like a big fence. When the tide goes out it will be completely dry in an hour and you will get everything that goes back in the mangroves. While the tide is out I pick up all the loose fish that are on the mud and clean the fish out of the nets. When the tide comes back in I pull the net back in the boat. I have caught many fish this way. The mud is slick and boggy, but one man can push a skiff over the mud and pick up all the loose fish. The pink spoonbills feed up to five and six feet of the boat and follow the tide out. They also feed at night. Always the catfish and crabs hit the net first, and then the mullet swim up and down trying to get out until the tide is all out, then most of them will go into the net or just die on the mud. We sleep while we wait. Mosquitoes feed first, then crabs, then catfish, then mullet and other fish.

George and I have done a lot of fishing down in the Chokoloskee area and all over that part of the country. One of our favorite places for silver mullet was Chevelier Bay, where we have caught many thousands of pounds. When we fished for silvers it was an all-night siege. We would leave our camp at Chokoloskee about four o'clock. It was about a two-hour run. We would leave down there or quit fishing when the morning star rose and head for the fishhouse.

THIS IS A DRY STOP FOR MULLET.
WAITING FOR THE TIDE TO FALL
IT WILL GO COPLEETELY DRY IN ONE MORE HR,,
WE ARE ASLEEP — IT IS AOUT SUN UP.

MOSQUITOS FIRST
THEN CRABS THEN
CATFISH THEN
MLLET & OTHER
FISH.

THESE PINK CURLEW, ARE
FEEDING. THEY ARE FOLLOWING
THE TIDE OUT. THEY ALSO FEED
AT NIGHT. OH WHAT FUN WE HAD

Dry stop fishing for mullet. Waiting for the tide to fall.

Mullet fishing in Pumpkin Key in the rowboat days [1934] was hard work. But I found an easier way. Fish didn't have a chance. I told Lem on one fishing trip that when we get in I was going to invent something that was going to be beneficial to fishermen or ruin a good skiff. I designed an inboard-outboard [1945]—a wonderful treat to fishermen. By cutting out the center of the skiff and putting the motor in the middle, the net can run out itself. All mullet fishermen soon had one. I didn't get a patent. But all the other fishermen and I have sure enjoyed it. Oars and oar locks are not used anymore. It helped the fishermen, but not the fish. It helped to make fish harder to find. Fish will not tolerate traffic. The slightest wake of a boat will run mullet off the feeding places or banks. Some banks are no good anymore—they are run over every day by every fisherman and sportsman that passes that way.

Fishing the old way, in Pumpkin Bay.

Fishing the new way.

Fishhouses in the area handled thousands of fish. While mackerel and kingfish were especially popular, mullet were the most plentiful. In December, during the run season, the fishhouses took in great quantities of mullet. Naples and Little Marco became the best places to catch mullet during run season. Ever-increasing boat traffic eventually made Naples less desirable. Mullet were caught mostly at night, and the crack of dawn was an especially good time to make a good catch.

In 1916 there were schools of mullet that reached from Gordon's Pass to the Naples Pier. Early in 1900 mullet sold for one cent a pound. By the 1940s it was up to six cents, and it had risen to twenty-five cents in 1976. Cotton nets were soon replaced by monofilament. In a thirty-five-year period, pompano went from nine cents a pound to $2.35.

I quickly learned every cove in the bay. At night I'd hear the mullet up in a cove and couldn't find a way to them. So I'd go back in the daytime and walk through the mangroves and find a small bay, cut a hole with a hatchet or potato rake big enough for the skiff, then I'd name the hole. Britches Hole was a favorite. Hewitt McGill and my brother Hub passed there once while I was fishing with my older brother George. We put our light out so they wouldn't find us. The next day Hewitt decided to find the spot, and when he did he put up a pair of cutoff dungarees over a mangrove tree so he could find it later. I named it Britches Hole.

I named First National because it paid off so well. Other holes were Cow Pen, Pennywinkle Hole, Whiskey Creek, Twenty-four Hundred Hole, because I caught twenty-four hundred pounds there once.[10] We had lots of homemade or secret places that no one knew, only the first finders. This is very interesting to a mullet fisherman, to find a new place that no one has ever been in.

We fished on neap tides [the tide occurring just after the first and third quarters of the lunar month] when the tide didn't fall very much, and on spring tides we fished the rivers in the upper bays. Pumpkin Key River was a long narrow creek that had some bays up at the head of it. It would take a couple of hours to row up there. We would start early in the evening so as to make it about dark for fishing. We would make a few strikes up there, and the fish we didn't catch would go on out and while we were coming out of this narrow creek these big mullet would be jumping in our boats. Mullet skip through the water, sometimes making three leaps and jumping twenty feet. We had a carbide headlight to see our way through the creek, and it would scare the mullet and sometimes one would hit you in the stomach and knock the breath out of you.

In the past we had a closed season on mullet from November 15 to January 20. But for the last thirty years there have been no closed seasons. Millions of eggs have been destroyed, and now the mullet are almost gone compared to what they use to be. I have seen schools a fourth of a mile long, and lots of these schools in run season. But not anymore.

Naples pier and post office, 1918.

Post office and pier as I remember it 1921   Naples P.

Unloading and cleaning a mullet catch at Combs Fishhouse in Naples.

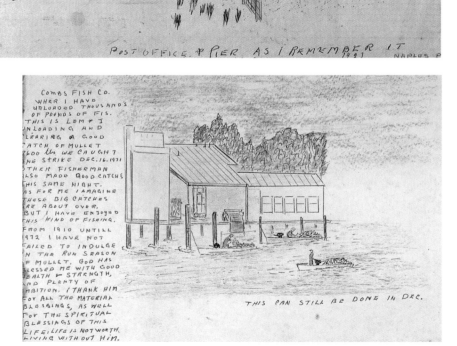

COMBS FISH CO.
WHRR I HAVE
UNLOADED THOUSANDS
OF POUNDS OF FIS.
THIS IS LEM + I
UNLOADING AND
CLEARING A GOOD
CATCH OF MULLET
2600 lb. WE CAUGHT
ONE STRIKE DEC. 16. 1971
OTHER FISHERMAN
ALSO MADE GOOD CATCHS
THIS SAME NIGHT.
AS FOR ME I AMAGINE
THESE BIG CATCHES
ARE ABOUT OVER,
BUT I HAVE ENJOYED
THIS KIND OF FISHING.
FROM 1810 UNTILL
1972 I HAVE NOT
FAILED TO INDULGE
IN THE RUN SEASON
OF MULLET. GOD HAS
BLESSED ME WITH GOOD
HEALTH + STRENGTH,
AND PLENTY OF
AMBITION. I THANK HIM
FOR ALL THE MATERIAL
BLESSINGS, AS WELL
FOR THE SPIRITUAL
BLESSINGS OF THIS
LIFE; LIFE IS NOT WORTH.
LIVING WITH OUT HIM.

THIS PAN STILL BE DONE IN DEC.

I have had some real disappointments in fishing. One time a conveyor belt in the fishhouse broke and 120 pounds of fish in a steel basket fell on my brother George's head and knocked him unconscious. I grabbed his head and cradled it, praying. I thought it had killed him, but I prayed and prayed and knew it was God or nothing. Soon George came to and said to put him in the cabin on the bunk. So I did and prayed some more. The ambulance came and George refused to go. He never fished again after that, but he lived.

About the last catch that George and I made was a real big catch of run mullet in New Pass Bay. About nine o'clock one night we loaded our boats so heavy with fish, I think we must have had about eight thousand pounds and we couldn't get out. George was sick, but we worked all night looking for the tide to come in and float us out. But it fell all night and all the next day and the whole place went dry, but we got our nets clean of the fish and it was a real hot night. The next day our fish spoiled. We couldn't get out to where our ice was. It was a real disappointment to us—the loss of money we would have made and the real hard work and our groceries were out at the launch with the ice. Well, when we got out to the boat just at dark that evening after all this we agreed to just forget that we had ever made this catch and not even mention it to each other anymore. But it was sure hard to forget.

George and I had another disappointment another night. Gordon's Pass had been full of fish at the Keewaydin Rocks at the mouth of the pass for several days. They wouldn't go out to where you could catch them. Other fisherman would watch them day after day. One night about midnight we decided to go and look for them again. We went, and they were gone. George was tired. He lay down for a nap. I went on looking for them. I went almost up to the pier, but couldn't locate them. Then I went south just off Keewaydin Club, and there they were, a black string about eight to ten feet wide as far down the beach as I could see. I went back to the boat and got George up and told him I had them located. We had this seine on the big boat and we went after them. I had George get out with the end of the net with a good anchor on it. He buried the anchor in the sand and I ran the net around them full length and took them in. We started pulling on the

net, and it was so hard to pull with all the mullet against it. But we kept pulling and it got easier. I said, "George, did you fix that anchor on the net to where they couldn't drag it off?" He said he did. I told him I was going down to the other end and check—for something had happened, the fish were not in there like they were. I ran to the other end—I had a good strong headlight on my head. When I got to the other end I couldn't see the end of the net. I waded out and couldn't even find it. When I got back to where George was the fish had all got away. All we had was an empty net. I suppose we lost a hundred thousand pounds. Now, that was really a disappointment. If I had just taken in a small portion of the school and secured the let-go end and buried the anchor real deep we could have caught more fish than we had boats to carry them. That was our last seine experience.

Coming in from New Pass Bay with mullet.

I have fished every season since 1910. When I started fishing in 1909 there were lots of fish. Mullet were *big* mullet. I have seen them go up the Everglade River in schools in the summertime, school after school. They would fill all the inner bays at night, and the next morning they would start coming out about sunup, and that was a sight to see, them passing our house which was right on the river bank. It would look like a hundred in the air at a time as they were going out in the bay. They would pass by for about an hour. When they got out in the bay they would scatter out and fill the whole bay. And this was not in run season. In the run season they would run at different places in big schools such as you never saw. They would start running about November 15. Now it is about the sixth of December, and there are not as many schools and no big schools. About the first of January it is all over.

If a shark or tarpon gets into a school it sounds like a shotgun going off, with mullet jumping everywhere. When they are scared and start running they can outrun a boat. I have handled lots of mullet and know what it was like and what it is now. In the old days they were salted and sold at Key West and shipped to Cuba. The fish companies, such as the Punta Gorda Fish Company, the West Coast Fish Company, and Riggs Fish Company at Fort Myers had fishhouses and a care-taker at most every place along the coast and boats to haul them. I can remember the first one that came into Chokoloskee Bay. My dad owned it and ran it. It would come in loaded with ice and go out loaded with fish. We used cotton nets, later linen, then nylon, and now monofilament. We used Spain cork for floats, which wore out or rotted and were easy to break; now we use plastic or sponge rubber. And we had to use limewater on the [cloth] nets every time you used them, and spread and dry them or they would rot. Now all you have to do is just fish them, leave them on the boat and forget that extra time.

In the early days it was hard to make fish hit your net in daytime on mud bottoms. But on grass bottom they would hit good. Now with monofilament nets you catch fish any place you find them. But it looks like the fish are about gone. The old grass banks that we would strike every day and catch fish, now you never see a mullet there because a boat swell runs over it every minute in the day.

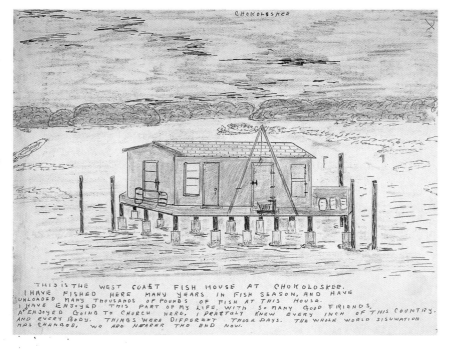

Unloading mullet at the West Coast Fishhouse.

The *Nanna*, a forty-four-foot cabin cruiser, was my first guiding boat. Before I owned the *Nanna* I guided Mr. Sillman from Wilmington, Delaware, and ran her for him. Henry Sillman was married to one of the DuPont girls and had a house on Gordon's Drive [Port Royal]. He put me on salary and I took them tarpon fishing during the winter. My sister Winnie worked as caretaker for them, too. One time I had the carburetors worked on and the mechanic didn't fasten the gas line tight. I didn't know it and thought everything was all right. But the next day on the way out to the grouper bars, about twelve miles out, I could smell fresh gas. When I stopped the motor I looked right quick and saw the gas spitting out. I yelled, "No smoking! We have a bad leak and the bilge is full of raw gas!" I bailed out over ten gallons of pure gas and mopped up the rest. I took up the floorboards and let them thaw out for a few hours before I cranked up again. It was a miracle we were not all blown up.

I purchased the *Nanna* during the latter half of the Second World War. I took Henry up Shark River, to Marathon, and other places throughout the Ten Thousand Islands. When the war started Mr. Sillman had joined the Navy and let the Coast Guard use the *Nanna* to scout for German submarines on the east coast, but by the time they got done with her she had been in a fire and the two new 130-horsepower Chriscraft motors were rusty and the boat looked destined for the graveyard. The Coast Guard had just towed her back to Naples and tied her up—gasoline was hard to get. I wrote Mr. Sillman expressing an interest in making a commercial vessel out of her. When my offer of three hundred dollars was accepted I pulled her out of the water, and my brother Wilbur and I cut her in two, right down the middle, and lengthened her by five feet and made a mackerel fishing boat out of her. I junked the motors and bought a new Chrysler motor.

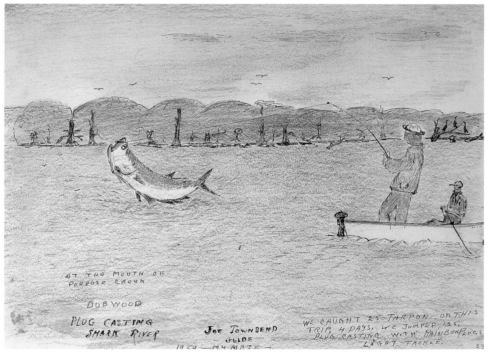

Catching tarpons.

Hunting, Fishing, Nature 99

'On the way to Shark River with a fishing party.

The first day I ever guided I remember the two old gentlemen, Buchanan and Birdseye. I got three dollars for that day and had a three-horsepower motor. We jumped a big 125- or 130-pound tarpon but didn't catch him. We threw the bait out when he jumped. This was at the mouth of Chokoloskee Pass, about 1909.

During my guiding years there were more jobs than I could accept. Florence Haldeman Price and her husband were regular tarpon catchers, although they never landed them.[11] They would pop the line when the fish had given them a good fight. Mr. Price sat on the boat and read Zane Grey books while they waited for a bite.[12] It was not uncommon for the Prices to hook as many as twenty-five tarpons with light tackle right at the mouth of Shark River.

One guiding trip was a sad reminder to me of how the world was changing. A couple from New York, the Wylies, and their two boys hired me as their guide. One of the boys' sport was to catch and kill things, especially by harpoon. We harpooned stingrays, whiparays, shark, tarpon, sawfish, and everything else that was possible. One day he harpooned a sawfish and wanted to harpoon a porpoise. This kind of went against the grain

The backyard of the Bruce Haldeman home in Naples.

with me and the other guide, Jack Daniels, who was one of the best guides in the country. But we gave in.

I had a real fast boat and could run a porpoise down in a few minutes. One day in Tarpon Bay in the Shark River area we found one, but the boy just couldn't seem to hit him with the harpoon. Finally he told Jack to try. Well, Jack got up on the bow of the boat. I put him right up to the porpoise and he stuck the harpoon in him. We thought we were really going to see some commotion, but when that porpoise found out that he was locked to some-thing he wouldn't do anything but just slowly swim along. Well, the big problem now was what to do with him. The boy started shooting him with a 22-caliber rifle, and every time he would hit him the blood would spurt out. He must have shot him twenty times before he hit him in the temple and killed him. Mrs. Wylie had her head covered up and would not look. I knew how she felt. I felt the same way. Well, it was all over, and we towed him to a shallow place and got the harpoon out. Jack said to me, "Rob, I'm glad this is all over. I feel like we have committed murder and I will never do this again." I told him I felt the same way.

Well, this was a sad lesson for us. I didn't approve, but didn't say anything. Every trip the boy would shoot at anything that moved—cormorants, pelicans, herons, ducks—anything that would make a target, and his dad was continually getting after him. I refused to take them anymore. Once George and I got a porpoise fouled up in our pompano net with the cork line around his tail and he couldn't get his head up to get air and he drowned before we could help him. I sure hated that.

"TUSSOK KEY IS GONE NOW, WASHED AWAY." ROB

One night after nine o'clock up Shark River in Tarpon Bay while we were anchored at Tussok Key a great comet appeared in the northern sky, for one night only. I think this was about 1960. There were twelve of us on the boat, and everyone saw it. The party was eating dinner down in the dining room of the boat, the *Polaris,* a big yacht my brother Wes was captain of. Joe Townsend and I were out on deck on the front end of the boat looking north, and there appeared a large comet, the biggest and brightest one that I ever saw. We called the party out on deck and they were so fascinated and surprised to see this big bright comet. No one knew anything about it. We never saw it again. There was no radio or television broadcast about it before or after. I have seen several comets; I saw Halley's Comet twice in my lifetime, but none as big and bright as this one. It appeared just a little west of the North Star. Most of the ones that I have seen appeared in the east and moved over to the west just as the moon does. Halley's Comet would rise in the morning like the daystar and you could see it in the daytime. It showed up very plain in the afternoon. I don't remember the year I first saw Halley's. I was a boy, but an old man that lived with us then said he saw it when he was just a boy.[13]

Tussok Key is gone now, washed away.

"SUDDENLY THE TIDE CHANGED, WHICH SHOULD NOT HAVE HAPPENED FOR TWO HOURS YET." ROB

One night my two brothers George and Hub and I were mullet fishing at Lostman's River. At the mouth of the river were lots of oyster bars, and the outgoing tide ran so hard you could hardly row against it. We rowed as hard as we could to get through the bars to fish for mullet in what was called Willie Bethel Cove. Just as we were about to make a strike for mullet, after such a struggle against the tide, suddenly the tide changed, which should not have happened for two hours. The moon was two hours high. George over on the right was in good fish. He hollered to Hub and me to let go. We were also in fish. But I noticed our boats were drifting very fast. We hesitated a minute before we ran any net. George had let

go and started running out his net. He hollered to us to let go, and wanted to know what were we waiting on. I hollered back that something had gone wrong, the tide had turned in and was running strong, that he'd better stop and see what was happening. By that time he had hundred yards out and it rolled up like a rope. In a few minutes it turned out again and ran out for two hours. This was a mystery to us. Never had such a thing happened. Old Brother Henry Smith told me the same thing happened to him many years before at this same place. I have experienced lots of strange things, and this was one I sure didn't understand.

Another mystery to me is about Naples Bay. When Acqualine Shores was being developed

with a dredge, there were pine stumps and cabbage stumps that were fastened together down where the dredge was digging, about four feet under the water. They had to put a wire cable on them and drag them out with a bulldozer.

Forrest Walker told me he had a pasture and a few cattle down at Myrtle Cove. He had a water hole that he had dug with a dragline for the cattle to drink water out of. One really dry season it almost went dry and he decided to dig it deeper and there were some fish in it, a little tarpon. The place had never overflowed and was a long long ways from the open saltwater.

"NET FISHING IS ABOUT A THING OF THE PAST." ROB

Dawn is beautiful at Turkey Key. When the shallow cover goes dry at low tide pink curlews feed. They follow the tide out. I have had them feed up to a few feet of me when I was fishing. Working a dry stop at one time between Marco and Goodland at low tide I could see hundreds of them feeding on the mudflats. Now you hardly ever see one in this area. Traffic has changed lots of things, especially fish grounds on grass flats near the channels

where mullet fed every day. With a gill net you could strike and catch mullet every day when the tide was right. But now a boat swell is rolling over it every minute of the day and they are not there anymore. Not many places are left for this kind of fishing, and what few places there still are where you might catch some mullet, there is someone there trying to catch fish—any kind of fish. Net fishing is about a thing of the past, especially on the inside route below Naples. Below Goodland is

the National Park and it's restricted to commercial fishing and they are feeling the results. I'm sad again when I think of it.

Many birds seem to be a thing of the past. Whooping cranes have almost been killed out. They are good to eat. One time George and I went on a fishing trip. We would trade fish for anything we could eat or resell, for pork, coon, hides, potatoes. This trip was between Venus and LaBelle. I saw a whooping crane

Whooping cranes.

A territorial great white heron.

Eagle perched in dead mangrove tree.

about three hundred yards from the road. I had a little single shot .22 rifle. I said to George, I think I'll try a shot at him. I didn't think I could kill him. I aimed high above him and shot and killed him. These cranes are hard to slip up on. They will always see or hear you and start whooping.

When I first came to Naples I did a lot of hunting. There was a crane that fed at a certain place every day. I tried to slip up on him to where I could get a shot at him but never succeeded. He always heard or saw me before I saw him. I only killed one.

In a tall pine tree near the Gordon River bridge was a bald eagle's nest. I saw it many years, and many baby eagles were hatched there. There was another one on the east side of Naples Bay near Haldeman Creek. Later years near Gordon's Pass there was an eagle that would be in a black mangrove tree every afternoon at four o'clock. The people from the camp across the pass from this tree would feed the eagle, and for many years he would be there for his meal at exactly four o'clock. I saw him many times as I passed by. He used the same tree.

Redheaded woodpeckers are about a thing of the past. I remember when there were lots of them. At Everglade when I was a boy there were hundreds of them. Along the dike between our home and store was a ditch and a lot of dead cabbage trees, and the woodpeckers had holes in about every tree and would raise dozens of baby woodpeckers every year.

There were lots of chicken hawks that often stole our chickens. When one was around for some reason he would always holler before making an attack. I would get the shotgun when I heard one and try to kill him. Sometimes I was too late, but I have killed lots of them to save our chickens.

Crows used to nest in a tall black mangrove tree just in back of our house and raised baby crows every year. They were so high up in the tree I couldn't climb up there to get one. I've heard a baby crow would make a good pet and could learn to talk if you split his tongue.

"THE MULLET ARE DEAD AND DYING." ROB

One afternoon during Christmas 1968 off Keewaydin Cove I caught about 3,500 pounds of mullet and left the rest of the fish in the net for Bem, my nephew.[14] My granddaughter Billie and her husband, Mokey, and Tiney, my dog, were in on this. We left about two thousand pounds for Bem. We almost sank before getting to a safe place to unload. It was Mokey's first experience with net fishing and he said he hoped his last.

In 1982 we had a real sorry run season for mullet. Not many big catches were made.

Rob and Tiney pulling in the net.

Bem and I didn't hit a good school. Our biggest strike was about 1,700 pounds. We had two such catches. We were right on the run every morning. About January 15, it looked like it was all over for gill nets. Seines made a few hauls. The best was 6,000 pounds for Nick Vanderbilt. Red tide was on. January 24 through February 8—it was still on. The mullet are dead and dying. Red tide seems to be moving south. Pompano fishing was good December 1981. Bem did good. I didn't do any good; in fact, I didn't try. Bem and I and Russ made a quick trip to Bahia Honda [near Marathon and the Seven Mile Bridge]. We had bad weather—only got one night of good fishing. That was good. Hub made a trip with Bem's boat on February 1. He had windy weather—only got two or three good nights out of nine. He did good. Now it is full moon and no mullet. Red tide is on again. The first red tide I saw was in 1917—everyone thought the Germans had done something to the water. There are south winds and showers. Cloudy, but warm. February 8, there are no pompano here yet. May 15 and we still didn't have any pompano. No silvers [mullet]. You can hardly catch a mess of mullet. Net fishing is poor. Very poor. A freak hurricane. Prayer turned it back twenty-five miles one night and tore it all to pieces. Now another one. Lots of rain June 18. October 9, Marilea went to Fort Myers for an eye implant [for cataracts]. Marguerite had one last week. I had mine last of July.[15] We have had lots of rain this summer—the most rain in many summers. October, so far, rain most every day. Some hard ones. I am ready to go fishing.[16] I haven't been in a long time. Only twice this summer for some to eat—once at Seagate, once at the Cow Pen. I haven't heard of many pompanos at Naples yet. Lots of pompanos caught from Fort Myers Beach with deep nets.

Marathon Seven Mile Bridge.

"IT WAS ALWAYS ONE MORE STRIKE WITH HIM, BUT I KNEW THIS TIME IT HAD ABOUT LEFT HIM." BILL SAVIDGE

Up until shortly before his death Rob fished, mostly for a "mess," or enough for a few meals. Knowing time was running out, his son-in-law Bill had retired from the city as public works director a bit earlier than planned in order to fish with him. Bill recalled their last run season.

The mullet were strung out along the beach and jumping, the gulf was like glass, the sky beautiful, and there were no boats around. Soon the school came together at Hurricane's Pass, porpoises got in and scattered them, and pelicans started diving for them. But they came together again. I thought how beautiful everything was. It was so perfect, and I hoped and prayed that the fish would strike. I knew it was his last time. And we ran out the big mesh net and the fish started striking. We caught nine hundred pounds. He had always been able to outwork anybody, but this time he sat down by the net and flipped the fish to me to take out of the net. It was always one more strike with him, but I knew this time it had about left him. I was thankful for this one last run—it had been a perfect strike in every way.

Rob confessed that not long before that last strike he had caught a netful of catfish and the energy wasn't there to clear it. He took the net full of fish up in a back bay and just left it. It was not unheard of to leave a catch.

"I SAW THEM TAKE IT ALIVE AND KILL IT, BLEED IT, AND IN A FEW MINUTES BUTCHER IT." ROB

Today, May 1, 1986, I am looking forward to a new experience. Earl [Rob's grandson] asked me the other day if I would like to go on a hog hunting expedition in LaBelle.[17] Soon he will get his [video]camera repaired and he can get some moving pictures. I think Vance [Cassie and Rob's son], Marvin Williams, Grant Thigpen, and his brother Bill and I are in the plan. Well, I have never been on one of these hog hunts, but I have tried most everything else. Maybe I can watch them and listen to the old hog hunt stories, which I'm sure they will have lots of them to tell while sitting around the old campfire. There is a saying, "Old dogs never get too old to learn new tricks." Maybe I can learn something out of this—if nothing else, then how good a soft bed will be when I get back.

[Later:] Well, the hog hunt was made. Only we didn't get to spend the night. We left home before daylight and got to LaBelle, where we were all to meet at the LaBelle Lumber Supply Company. We all met before daylight. We had to wait on Bill Thigpen, as he was a little late: he and his friend stopped at the convenience store on the way for coffee and donuts. But just at daylight we all got started—Vance in his Jeep, Marvin with his Jeep in tow, Greg Thigpen with his buggy and dogs. Bill and his friend were in a truck. Earl and I were in a car. We started out for Vance's camp or the ranch.

We arrived early and started right off for a hog hunt. I was with Vance. Earl was with Greg on a high seat in the buggy to take the pictures. Bill and his friend were with Marvin and his dogs on Marvin's Jeep. Not far from the camp the dogs found a hog—a red boar hog. They soon tied him up and had him in the Jeep and were off for another. We hunted quite a while, but didn't find any for a while. We saw two deer. It was about eleven o'clock by this time. Vance said he had a trap "over yonder" and we would go check it. So on we went. And there were two in it, two sows. One was about ready to have piglets. The other was a sow about just right to eat. So they killed this one and let the other loose. They took the red boar out of the Jeep and cut him and put him in the trap pen.[18] And we went on to the camp and butchered the one and cut her all up and salted and peppered her and got ready to barbecue her for dinner. We went hunting for another hour or so and caught a black sow ready to have piglets. We turned her loose and took the dogs and went to the camp. On the way in we cut some swamp cabbage. Vance cooked the cabbage. Marvin barbecued the hog, and by then everybody was starving. When it was all ready Bill asked the blessing and we went at it like a bunch of starving hogs.

Soon we were on the way for another hunt. We saw two more deer—only a flash and they were gone. It was so thick with myrtle you couldn't get a picture or a shot of them. Well, we didn't have any luck on this round, so we came back by the trap to see how the cut boar was, to turn him loose, and he was dead and stiff. We towed him off a ways to where the buzzards could eat him.

Well, this was all really an exciting trip for me. Those boys really knew how to do this, as they had done this all their lives. Of course, we had some good hog stories while we were at the camp preparing the meal. The movie film was just fine. We came back by Greg's home and showed the film on his television and it was good.

Well, I enjoyed this hog hunt so much. I hope we can have another one. I would like to help plan (and cook the cabbage); Marvin

A mad wild boar.

[can do] the drinks—Coke and tea or whatever, Grant the bread. Bill will bring the blessing. And I believe that will take care of everything. If we kill a hog we can bring it back for our wives, as they can't be with us at this good meal and fun and hear all those old stories.

Just a little more about the hog hunt. I didn't want to mention this, but I didn't enjoy the fresh barbecued pork. I saw them take it alive and kill it, bleed it, and in a few minutes butcher it and salt and pepper it while it was still almost alive and in about two hours they were eating it. Well, this was all new to me, and after watching the whole performance I just didn't enjoy it. But the old-timers did. And this is why I want to help plan the food. I know they will all enjoy a fish fry just as much as a fresh pork barbecue. I guess I shouldn't have mentioned this, but that's just the way it was.

July 28, 1986. Another hog hunt and get-together for a good meal. It didn't plan out just like I had it planned (the fish dinner), as I couldn't get everything organized for the fish. So we didn't have fish, but plenty of good barbecued meat that was really good. We had our wives meet us at noon with lots of goodies and we really had a good get-together and a real good eat out. After the meal they went with us on a little hog hunt. We didn't get any hogs, but had a big black boar in a pen trap and cut him and turned him loose. It was very exciting—had everybody scared for a while when we turned him loose and he almost whipped the dogs.

When the wind is out of the west, the sheepshead bite the best.

When mangoes are ripe, the grouper will sure bite.

When the red mangrove is in bloom and the seeds are full, mullet fishing is good.

THIS HUNT TAKEN PLACE IN 1915 —
THE PRICE OF FISH DROPED FROM 2½ CENTS TO 2¢.
WE WENT ON A STRIKE, AND WENT HUNTING.
ON THE WAY UP THE RIVER WE KILLED ABOUT
12 CURLEW. THIS BIG IRON POT IS FULL — THEY
AM STEW FRYING, AND THE OTHER POT IS RICE,
AND THAT BIG DUTCH OVEN IS FULL OF SODA PINE BISCUITS.
AND A BIG POT OF MAXWELL HOUSE COFFEE. and 8 HUNGRY MEN
THAT CAN HARDLY WAIT FOR MR MATHEWS TO SAY COME AND GET IT.
THIS CAMP WAS THE ONLY DRY PLACE AT THE HEAD OF NEW RIVER
WAS AN OLD INDIAN SAND MOUND NO TREES NEAR. WE HAD TO
PICK UP SOME BUTTON WOOD COMING UP THE RIVER FOR FIRE.
IT IS ABOUT NIGHT NOW. WE'LL HUNT TOMORROW. ITS ABOUT A MILE
ON OUT TO THE CYPRESS THROGH ALMOST KNEE DEEP OF WATER
JOE KILLED ONE BIG BUCK — IT WAS JUST TO MUCH WATER .
BUT WE SURE HAD A GOOD TIME AND EJOYED GOOD EATING

JUST ONE SMELL OF THAT
CURLEW POT, AND THOSE GOOD
BISCUITS COOKING, AND A SENT
OF THAT MAXWELL HOUSE. WOULD
ALMOST CURE ANY STOMACH TROUBLE

WHAT A WONDERFUL NIGHT
LYING IN BED HEARING THOSE
BIG BULL FROGS. BRASS BAND.
PLAYING ALL NIGHT LONG.
ALL THESE FRIENDS ARE
GONE. BUT — ME —

NAMES
"MR JOHN MATHEWS"
CHIEF & COOK
GREGORY LOPEZ
JOE LOPEZ
ALPHONSO LOPEZ
HARY McGILL
WALTER HOWELL
LUCIOUS WATSON CO
ROB STARTER

WE LEFT OUR LAUNCHES
FARTHER DOWN THE RIVER
I AM THE ONLY ONE
LEFT OF THIS. TO TELL OF IT
1975

Memories of a good hunt in 1915.

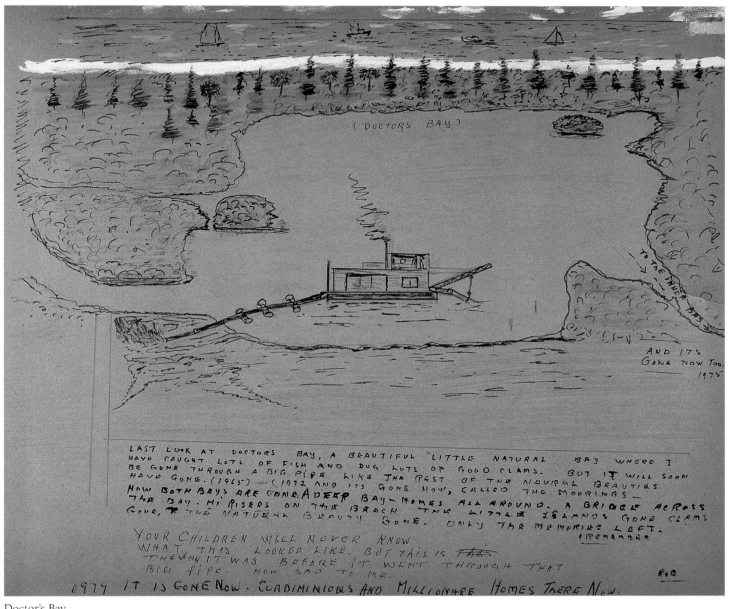

(DOCTORS BAY)

TO THE INNER BAY

AND ITS GONE NOW TOO. 1975

LAST LOOK AT DOCTORS BAY, A BEAUTIFUL LITTLE NATURAL BAY WHERE I HAVE CAUGHT LOTS OF FISH AND DUG LOTS OF GOOD CLAMS. BUT IT WILL SOON BE GONE THROUGH A BIG PIPE LIKE THE REST OF THE NATURAL BEAUTIES HAVE GONE. (1965) (1972 AND ITS GONE NOW, CALLED THE MOORINGS NOW BOTH BAYS ARE GONE A DEEP BAY-HOMES ALL AROUND. A BRIDGE ACROSS THE BAY. HI RISERS ON THE BEACH THE LITTLE ISLANDS GONE CLAMS GONE, THE NATURAL BEAUTY GONE. ONLY THE MEMORIES LEFT.
I REMEMBER

YOUR CHILDREN WILL NEVER KNOW WHAT THIS LOOKED LIKE. BUT THIS IS THIS THE WAY IT WAS BEFORE IT WENT THROUGH THAT BIG PIPE. HOW SAD TO ME.
P.O.

1979 IT IS GONE NOW. CORDIMINIONS AND MILLIONAIRE HOMES THERE NOW.

Doctor's Bay.

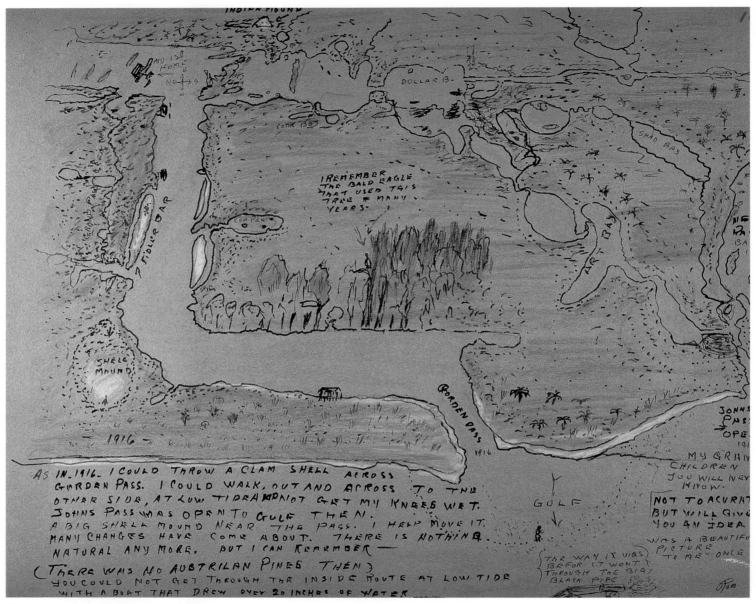

The map contains the following handwritten labels and text:

INDIAN MOUND

MY 1st HOME

DOLLAR B.

COON 1363

SHAD BAY

I REMEMBER
THE BALD EAGLE
THAT USED THIS
TREE # MANY
YEARS.

GRAND PASS

FIDLER BAR

COW PEN

OYTER BAY

SHELL
MOUND

GORDEN PASS

1916

1916

JOHNS
PASS
OPE
191

GULF

MY GRAN
CHILDREN
YOU WILL NEY
KNOW.

NOT TO ACURAT
BUT WILL GIVE
YOU AN IDEA.

WAS A BEAUTIFU
PICTURE
TO ME - ONCE

THE WAY IT WAS
BEFOR IT WENT
THROUGH THE BIG
BLACK PIPE

AS IN 1916. I COULD THROW A CLAM SHELL ACROSS
GORDEN PASS. I COULD WALK OUT AND ACROSS TO THE
OTHER SIDE, AT LOW TIDE AND NOT GET MY KNEES WET.
JOHNS PASS WAS OPEN TO GULF THEN.
A BIG SHELL MOUND NEAR THE PASS. I HELP MOVE IT.
MANY CHANGES HAVE COME ABOUT. THERE IS NOTHING
NATURAL ANY MORE. BUT I CAN REMEMBER —

(THERE WAS NO AUSTRILAN PINES THEN)
YOU COULD NOT GET THROUGH THE INSIDE ROUTE AT LOW TIDE
WITH A BOAT THAT DREW OVER 20 INCHES OF WATER

Big shell mound near Gordon's Pass.

SO MANY CHANGES

I WAS BORN ON NAPLES BEACH.
BUT NO TELLING WHERE I WILL
WIND UP — WHEN I CAME BACK,
THERE HAD BEEN SO MANY CHANGES! COULDENT
EVEN FIND WHERE I WAS BORN.

Blue turtle.

THIS IS A SHRIMP BOAT ——— MY BROTHER HUB DELIGHTS IN
THIS KIND OF FISHING — SAME AS I HAVE IN MULLET FISHING —
HE HAS BEEN VERY PROSPERS IN IN THIS BUISNES. HE HAS OWNED
LOTS OF THESE BOATS. AND HAD LOTS OF EXPEARINCES, HE HAS LOTS
OF GOOD STORIES OF THIS.
ALL I EVER WANTED OF THIS KIND
OF FISHING. WAS ALL THE SHRIMP I
COULD EAT.

Shrimp boat.

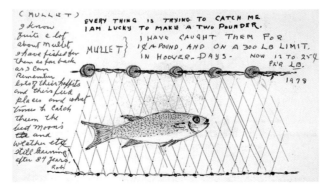

(MULLET)
I know
quite a lot
about mullet
I have fished for
them as for back
as I can
Remember
lots of their haffits
and their feed
places and what
times to catch
them the
best moon's
tide and
weather etc
still learning
after 84 years.
Rob

EVERY THING IS TRYING TO CATCH ME
I AM LUCKY TO MAKE A TWO POUNDER.

MULLET
I HAVE CAUGHT THEM FOR
1¢ A POUND, AND ON A 300 LB LIMIT,
IN HOOVER - DAYS. NOW 12 TO 25¢
PER LB.
1978

Mullet—Rob's favorite fish to catch.

The old mackerel fleet.

There are lots of different kinds of fish. I dont know them all. I have been fishing ever since 1909 and have handled lots of fish. many kinds of fish. I have lots of good fish Stories — some would be hard to believe, but They are True. I have fished all over the south west coast of Florida in all the Rivers and Bays and inland waters all over Bahunda some in the Fla keys and gulf Stream — I know the ten thousand Islands Very well — at one time I knew every body that lived on them or Something about them. Since it is a Park the people have been forced to leave and Vacate them. Can hardly camp on them any more without a permit. They are all grown up with wild Bushes + trees so that you would never know any one lived on them — many families were born + Raised on them — My grand childer will never know them as I did. The Park may seem preservitave to many people. But It makes me sad to see it now and think how it use to be. I may be an old fogy. But I still rember the Islands + People that lived on them and enjoyed life as it was in my early days. I am glad I lived in those days — I am 91 years old now and can still remember how it realy was in those days. Things are different now, as I suppose they are to be. but some things are hard to forget. well we are nearing the end — the last days — and will soon be over. and a new earth and a better one. I am looking forward to that time — where we will never grows old — and live for ever and ever ∞ ever. after a million years we will have no less to live.

PRAISE THE LORD.
- - - - - - - - - -
MY FRIEND THE PELICAL

A REAL FRIEND.
THEY FOLLOW ME AROUND
I GIVE THEM FISH TO EAT
SOME TIMES THEY HELP
THIR SELF TO FISH — OUT OF
JOURNET. OR BOAT.
HE HAS HELPED ME
FIND FISH THAT I WOULD
NOT HAVE FOUND WITH
OUT HIM.

Rob.

Pelicans love lady fish. Bill + I was out fishing the other day - and one was following us around. When we made a strike we got a lot of lady fish - and we would toss one out to him as long as he wanted them he ate 5 befor he Said he had enogh.

Dr. Earl Baum, Naples, 1938, with a twelve-hundred-pound manta ray harpooned by Dr. Baum and Forrest Walker, a local guide, neighbor, and good friend to Rob and Marilea. The manta ray had a wing spread of twelve feet and three inches.

Rob's fishing trip gear: yellow slickers and jacket, rain hat, rubber boots, lunch box and thermos of hot coffee, wooden "cat" club, lead staff, bail bucket, carbide light, carbide cap, oar locks, oars, water jug, carbide in glass jar, pocketknife, net rings, lead drag, lantern, cloth gloves, and mosquito lotion.

Preston Tuttle in the 1925 Stu-Tutt car he and Arthur Stewart built.

Lem Storter and his dog, Heidi, hunting deer in the Everglades.

1972
OUR
HOME IN ORTONA,
AND GEST HOUSE
A BEAUTIFUL PLACE TO LIVE.

Rob's beloved home in Coffee Mill Hammock.

SEVEN ➤ FINAL DAYS

I was just thinking today of the old times when I was a boy living in Everglade. I had twenty-two cousins. We went to school together. We were as one family. We often had birthday parties. How we loved that twenty-four-quart ice cream freezer (when we could get ice). All of these cousins are gone but two. How sad. All these cousins were dear to me. Along with these cousins there were other families of children that were dear to me, too. All are gone now, but I still remember.

Last night I dreamed I saw a big hole in the sky. I saw the second heaven. I saw temples high and lots of other things, I can't remember what they were.

I dreamed that Marilea, Betty, and I were in a boat moving along in smooth water. Marilea was steering the boat. She told me I had better take the control—that there was awful rough sea just ahead. But all at once we were right in to this real rough high sea, and before I could get up to the bow to take the wheel a real high breaker hit the boat and swept it, bow down, to the bottom—the stern was straight up. The boat almost broke up and almost sank. It looked like it was the end. I thought about Betty, how I would swim to the shore with her when I was almost helpless myself. But I hollered to Marilea to try and keep the boat up just a few minutes more. She did. She steered it right on through into calm still-water, which was a miracle. When we got into calm water, we talked. She said, "I warned you. I saw that danger just ahead." I said, "Well, I didn't know it was so close to us." We waited too long and the current carried us out, but with the Lord's help we made it. Now, just what this means I don't know, but it could mean several things. For a few minutes it was the "awfulest" thing I ever experienced in my whole life. It looked like the end of everything.

It is now August 2, 1986. Lem says mullet are not showing any fat yet. August 4: I see Bill getting his boat on tow, to go fishing today. I planted peas today (new moon). It seems like my fishing days are about over. My boat is pulled out and needs lots of repair. My motor is in the garage on the rack. And me, I'm out of pep. Well, I am still looking for good health again. Soon. I am standing on the promises. I still remember the old song, "Standing on the Promises I Cannot Fail."

I am sitting by my window just thinking, and watching the cars zip by, and now and then someone walking. I just saw an old man hobbling along. I suppose he was just out taking exercise. Well, I get lots of exercise too, most of mine is in the backyard with a hoe and rake and shovel and wheelbarrow—sometimes, a hammer and saw. Well, I am not complaining. There were lots of days when I had good health, strength, and energy and felt good, not bad. There is no machinery that gets so worn that the man who made it can't repair it again.

August 8: Lem brought me some fish today. I see a little fat started. Bill cleaned one with good fat. Each week they will be showing better fat.

At one time Naples was booming with tourists. The guides were busy every single day. There was lots of fish and lots of guides: Dan House, Cecil Lamb, Jack Cannon, Major

Regis, Salvador Gomez, Jim Hatcher, Buren "Boo" Davis, Henry Ernshaw, Eric Vickers, Preston Sawyer, Lee Parker, Hub Storter, and me. At times, Floyd O'Bannion, Lem Storter, Willie Tomlinson, Ernie Carroll. Most are gone now. This kind of guiding is not being done these days. Most everybody has his own rig. Anyway, the fish are not here like they once were.[1] In my early days of guiding out of Naples, any time that the gulf was smooth you could make a good catch of fish. If it was too rough you could do good fishing on the inside [in the bays].

My boat somehow sank at the dock. Things floated out—cans, oars, fish box, etc. The motor was completely underwater. Jim Gammell is overhauling the motor. He is good at that. Just what caused the boat to sink I don't know, unless there was too much rain and the drain hole in the back was underwater and the net was too heavy. The water came in from the back. The boat didn't leak. This is the third time that motor has been completely underwater at this dock.

We have it [the motor] all right now and ran it on the rack.

Marilea and I are working on a new doghouse now. We have three dogs, Tootsie, Bonnie, and baby Pannie. We sure hope they will appreciate it. It's hot, hard work.

The boat got caught under the dock at the stern and sank. The drain hole in back was open and caused it. I pulled the boat out. Had lots of help. Bill, Earl, Mokey, Big Jim Cain, and Peggy's Jim will probably repair it later.

Bem is down in Naples now. I don't know what for, unless he is spying around to move back to Naples. I thought he made a bad and regretful move.[2] Well, he has got the pep. He will soon be in shape again.

July 27, 1986: Tootsie died. I buried her in the back. I couldn't help but cry when I put her away. I guess she went where all the good dogs go.

We just spent a few days in Coffee Mill Hammock.[3] We did a lot of cleaning up. Coons are awful bad, they sure raided the pineapples.

LET'S GO TINY —
SHE WAS ALWAYS READY—
NEVER LATE—
TINY WENT FISHING WITH ME EVERY TIMEM, DAY OR NIGHT. I ALWAWS SHARED MY LUNCH. HER, SHE PRACTUALY NEW ALL THE CROKS AND TURNS IN GORDEN RIVER. ALWAYS WHEN COMING IN AT NIGHH SHE WOULD GET UP OUT OF BED WHEN I MADE THE LAST TURN SHE KNEW WE WAS NEAR THE DOCK. I MISS HER SO MOCH NOW. SHE HUNTED WITH ME, SHE KNEW LOTS DEER TRAILS THAT I HED LEARNEN HER. SHE ENJOYED HUNTING AS GOOD AS I DID.

Rob and Tiney, his constant companion.

They didn't even wait for them to get ripe. Time and tides wait for no man—coons and armadillos wait not for pineapples to get ripe. Next year we will have to pick them earlier. Tooke [Olivia], Bill, and I spent Friday and Saturday at Ortona. We did a lot of mowing. It looks good.

The first hurricane of 1986 (Bonnie) is off the Texas coast. Number two is started (Ethel), with winds over 75 miles per hour and twenty-foot waves near Hawaii.

A beautiful day. Kind of warm, a few white clouds scattered above. Who would complain? This beautiful day has turned into lightning and squalls, some rain.

Hurricane Charlie is off the coast of South Carolina. No threat so far. It has weakened down to just a storm. A woman on television said Charlie was a man, that's why it didn't do anything. Who will be next?

September 9: One small thirty-five-mile-per-hour storm way out in the Atlantic. Marilea, Tooke and Bill, and I spent a week in Ortona, picking peas and shelling them. Did some cleaning up, mowing, etc. My mower broke down. Lots of rain this month so far. Sold the van.

"I CARVED MY INITIALS ON THIS TREE." ROB

A true story of three little twigs: God made them for a purpose. They hung on a tree until they were ripe. They fell in the water and started drifting. They were about seven or eight inches long, sharp and a little larger on the bottom, and they drift in the water until they find a lodging place and they take root and make a tree. That is the way the Ten Thousand Islands were made. So these three little twigs fell in the water. They started drifting, and one got in the strong current and drifted out to sea—so far it never found any lodging place and was never seen again. Another one drifted onto a mudflat. It rooted and made a small tree, but a storm came and washed it away and it was never seen again. The third one drifted until it lodged on a coral rock and it rooted. Each year it put out more and more roots until it was well rooted, and up came a storm, but it didn't move it and it kept growing and every year it put out more and stronger roots, and another, bigger storm came and broke up some of its limbs, but it stood right there, and it stood through several storms, but they couldn't wash it away because it was rooted and grounded on the solid rock.

This story is true. I saw this tree when I was a boy in 1906. I carved my initials on this tree. And I passed by it hundreds of times and was amazed how it could stand so much. So it is with us. We are like little mangrove twigs and are drifting somewhere to find a lodging place. Let's hope we will not stop on the mudflat or drift out to sea, but will find the rock and there we will be rooted to stay. This Rock is Christ.

"I DON'T KNOW WHAT TO WRITE, MY WRITING DAYS ARE OVER, BUT I HAVE LEFT ENOUGH IF YOU CAN GET IT TOGETHER." ROB

Rob wrote this in his journal around Christmas, a few weeks before his death. He had gone into the hospital and had surgery the day before his ninety-second birthday, September 29, 1986. He woke up the day after his surgery and said, "I made it to another birthday." He spent the next few months at home trying to recuperate. But he never did.

Bill would put him in a chair and take him outside in his garden, Lem brought roses and fresh fish, and his sister Winnie brought fish chowder. Olivia, Marguerite, and Lem's wife, Lucy, became his caretakers. My twin sister, Barbara, and I were there in his final days, taking turns at rubbing his feet, holding his hands, and talking to him. Family came and went, and when Marilea realized his time had come she bent down over

him and cried, "Honey, I can't go no farther with you," and she could not bear to be in the room with him. As he took his final breath Barbara told him we would take care of Mema (the name we gave my grandmother when we were toddlers); Marguerite bent over him and told him she loved him, and Olivia whispered in his ear, "Daddy, Jesus loves you." And he was gone.

"WHEN THE ROLL IS CALLED UP YONDER I'LL BE THERE. PRECIOUS MEMORIES TO ALL." ROB

This was the last journal entry, on January 1, 1987. He died eleven days later.

NOTES

One. Early Days

1. A coon skiff is a small, narrow rowboat about eight to ten feet long, propelled by oars.
2. Todd, a Tennessee native, was appointed by the Monroe County School Board in Key West.
3. Today's Collier County, founded in 1923, is part of what was formerly Lee County, which in turn was created from Monroe County.
4. Fort Winder was a gathering point for federal troops in the campaign against the Seminole Indians.
5. Everglade was formerly called Allen's River and Potato Creek. Also living there was Madison's brother, John, a deserter from the Civil War. ("I would fight one man," he said, "but not a regiment.") Madison's son, Andrew, cleared hammockland for $3 a month and stuffed panther hides with grass and palmetto fronds for $55 each.
6. Plume hunters were hunters who killed the American and snowy egrets and roseate spoonbills for their plumes or feathers.
7. Buttonwood, a mangrove, was an important source of firewood and charcoal.
8. This claim was also made in the *Estero American Eagle* in 1912, and the *Fort Myers Press* stated that thirty acres of Storter cane produced nine thousand gallons of syrup.
9. R. B. Storter and Dick Myers earned $1.50 a day guiding said Mr. Wilson, who went back north and patented the first fishing spoon, the Wilson Spoon, a shiny curved lure made of metal.
10. This story may in fact have been about an incident in which Indians killed a white woman and her baby. Both versions have been told.
11. Fort Myers is perhaps best known as the winter home of Thomas A. Edison, who came there in 1886 in search of a suitable filament for the incandescent lamp he was later to perfect. It is said that after finishing work on the electric lamp he offered to install free lights in town, but the town rejected his offer because the lights might keep cattle awake.
12. Shell—ground or broken-up shells—was used for building roads, among other things.
13. Optimism of the 1920s brought two trains daily to Marco—for a time, anyway. The Great Depression contributed to the abandonment of the railroad around 1942.

Two. Daily Life

1. A sloop has one mast; a schooner has two, with the front mast shorter.
2. A smack was a small boat with a well for keeping fish alive.
3. Sofkee was coarse grits cooked for hours.
4. The men always ate first so they could go off hunting.
5. Edward Wotitzky was said to be "generous to a fault" and extended too much credit to make his business profitable.
6. Em Gardner later married Rob's grandfather, John Allie Stephens. She sold alligator hides for $1.25 for a seven-footer, and coon hides for anywhere from twenty-five to seventy-five cents.
7. Hookworms, parasitic roundworms that enter the body usually through bare feet, affect the small intestines, causing anemia and weakness. At the turn of the century South Florida had one of the worst rates of hookworm disease in the country.
8. Reference is to the writer A. W. Dimock, who was accompanied by his photographer son Julian. A. W. wrote *Florida Enchantments* in 1908.
9. On occasion, the Storters also hired Seminoles to help cut cane.
10. Rob did the same thing to his own boat, the *Nanna,* years later.
11. The fine for shooting a turkey was $25.
12. A "cypress head" is a cluster of dome-shaped cypress growing in a low area.

13. Millions of clams were processed into clam broth and chowder between 1903 and 1947 on Marco Island. A huge clam bed stretched forty miles through the Ten Thousand Islands. Preston Sawyer, one digger, remembered that he wore heavy canvas shoes. Still, the bottom was so sticky that the spurs almost ruined their feet. The clams were collected by means of a 110-foot-long dredge with an 800-pound anchor that was lowered into the shallow water; the barge was then allowed to drift to the end of a 1,200-foot cable. As the cable was drawn in, the dredge "digger" pulled up as many as 125 bushels of clams, which were put onto a conveyor belt just below the water to rinse the mud off. At the top of the belt four men sorted the clams into baskets. Sometimes the dredge operated twenty-four hours a day.

14. Although there was apparently little crime, the police had a reputation for brutality. Cale Jones, who succeeded his father as police chief, was known to go into McDonald's Quarters—the black housing district—and beat up troublemakers instead of arresting them. His reputation extended into the white community as well.

15. The first electricity was supplied by two diesel engines located in the ice plant. At the hotel, around ten o'clock, the lights would blink, signaling that the diesels were about to go out and candles would have to be lit.

16. Naples's earliest settlers were Roger Gordon and Madison Weeks. In 1885, when the newspaper publisher Walter Haldeman of Louisville, Kentucky, selected Naples as a winter home, he paid Weeks $400 for his holdings. Haldeman, General John S. Williams of Winchester, Kentucky, and several other gentlemen, under the name The Naples Company, bought land and charted the streets for a town. They sold lots for ten dollars. Haldeman owned so much land that for a while he gave 33½ feet with each subscription to his newspaper. Winter visitors in the early 1900s said there were "natives" in Naples "who told some very tall stories of hunting and fishing."

The canal, which extended a little over a mile across the peninsula, was recorded on a 1775 map and ran from the Gulf of Mexico at Ninth Avenue South to the Gordon River (Naples Bay) near Fourteenth Avenue South. The best guess is that it was dug by Indians by hand. The Naples Improvement Company had plans to restore the canal but never did.

Three. Family and Friends

1. Among the alternatives to tobacco proper was rabbit tobacco, the leaves of which were stripped off the bush, crushed, and rolled into cigarettes.

2. A "mosquito bar" is a tentlike structure, usually made of cheesecloth, hung over the bed by string to keep the bugs out.

3. La Brisa was a popular pavilion on lower Simonton Street, which was almost destroyed in the 1910 hurricane.

4. John Allie Stephens had more than one mail-order bride; all returned to their homes shortly after moving to the Ten Thousand Islands.

5. In fact, he was probably between 95 and 100 when he joined the church.

6. José Gaspar made Gasparilla Island (near Boca Grande) his headquarters. He is said to have wrapped an anchor chain about his waist and leaped into the sea in 1822 when his ship was overtaken by the *Stars and Stripes*. He kept a diary from 1784 to 1795 recording his captures and amount of plunder.

7. The South in this period had the highest infant mortality rate in the nation. Women were traditionally the caretakers at birth and death.

8. Fine shot is tiny pellets meant, for instance, for birds. It doesn't kill a large animal but might put an eye out.

9. The cabbage palm or sabal palmetto, the Florida state tree, is the source of swamp cabbage, often eaten by the pioneers. The bud, or heart, is eaten raw or cooked, often with salt pork or white bacon. Harvesting swamp cabbage kills the tree.

10. Married women were duty bound to be available to their husbands, and while men had great latitude in fulfilling sexual desires, women were expected to suppress desires. Marilea said when the boys came in from fishing one of the first things they wanted after they ate was to "get to their wives." The blanket separating their sleeping rooms had a hole in it, and one time Zola, her brother Charlie's wife, was visiting and peeped through it, then later bragged about it. Marilea said she didn't have much use for Zola after that.

11. Although little medicine was available to treat illness or mineral and vitamin deficiencies, home remedies were common.

12. Rob and Marilea even tried taking Betty to the Oklahoma evangelist Oral Roberts, who was holding a tent revival in St. Petersburg.

13. Census records from the early 1900s tell of the hardships many women endured. It was not uncommon for women to marry in their early teens and have children well into their forties. Daniel House, twenty-five, married a fifteen-year-old—only six of their nine children survived. Old Henry Smith took a twelve-year-old bride, who had her first baby at fourteen. Thirty-eight-year-old John Tomlinson's wife, Mercedes, had children aged ten, nine, eight, six, three-year-old twins—but only one lived—a two-year-old and a two-month-old.

Four. Faith and Religion

1. Chokoloskee, where McKinney's Landing was located, is a tiny shallow island about ten miles long and less than two miles wide. John Weeks was the first white man to live on the island, having arrived there in 1870. When he moved, the Santini family claimed the island; others, like Mr. Carroll, the lighthouse keeper at Key West, visited but did not live there. C. G. McKinney, the first medical man and post office owner; Ted Smallwood, the trading post owner; and Dan House, the principal landowner, followed with their families. Dan owned the *Rosina,* a fifty-two-foot schooner that made trips to Key West. His wife sold grasshoppers to the American Entomological Company in Brooklyn, New York, for which she was paid one and a quarter cents for each grasshopper.

2. Hub Storter confirmed the miraculous changes that happened overnight in some of the converts. "Old coon hunter Tant Jenkins got salvation and died a good man, but before that he would go out and hunt a whole season and come in with a boat loaded with coon and otter hides, which he'd get about a thousand dollars for. Then he'd spend every bit of it on drinking. Jack Cannon was the same way, and he became a devout Christian. When Jack wasn't going to church, his wife was; then when Jack changed and started going to church, his wife stopped."

3. The Lopez family was Catholic and was regularly visited by traveling priests.

4. Nora Cassie Williams had married James L. Williams when she was twelve years old.

5. It was never determined what killed Cassie—she had been healthy up until the night of her death; however, typhoid, being an acute infectious disease acquired by contaminated food or water, seems the likely cause.

6. A victory march around the building or altar with everyone singing a hymn, fashioned after the biblical march around the walls of Jericho.

7. Cassie's parents lived in Naples awhile, then in Fort Center, where they bought hogs from the Indians; they eventually settled in Ortona. Vance became a Glades County commissioner and rancher.

8. Rob and Marilea continued visiting Hen Scratch—camp meetings were the most recurring social event at that time. The first thing the kids would do was pull Spanish moss off the trees and spread quilts over it to make beds. Church lasted from daylight to dark.

Five. Hard Times

1. Belle and her husband, Sam, a Cherokee, had an outlaw band in Arkansas. When Belle threatened to turn Watson in to the authorities in Florida, he killed her. A prominent Fort Smith attorney got him off, and he eventually fled back to Florida.

2. Claude was told that Watson had returned to the island after Smith and Waller had been killed and helped kill Dutchy.

3. It has been reported that Mrs. Smallwood, knowing her husband had been threatened by Watson, had sold Watson shells that had gotten wet in the storm.

4. It is generally agreed that Henry Short (Shorty), who worked for the Storters, was the first to fire on Watson. The 1900 census shows Henry, a hunter, to be a single Florida-born black man, twenty-eight years old, unemployed five months of the year.

5. The child was reportedly Ed Watson's infant girl, who was swept away while her mother, Josie Jenkins, sought refuge in a mangrove tree.

6. R. B., Claude, and the Martins spent the night on Fakahatchee Island in total darkness after their fire was accidentally extinguished. They placed sticks in the ground to determine whether the water was still rising. When the water had almost completely covered the island, it began to recede: the worst of the storm had passed. It took thirty-six hours for R. B. to get back to Everglade, where he found his family safe, but his collection of books, including first editions and encyclopedias, had been ruined.

7. The *Speedwell* was hit by a squall about eighteen miles from Key West. Nine passengers were drowned in the cabin: the captain's three sons and six members of the Bradley Nichols family from Bridgeport, Connecticut.

8. Dan House, Lemuel Storter's father-in-law, was a Chokoloskee rumrunner for thirteen years. At first he ran the boat himself, carrying a gun to ward off hijackers. His children remember helping to hide liquor in the attic, storing sometimes two hundred cases in one night. He would unload the liquor at the head of Gordon River. An old man who lived next door to them let Dan store the whiskey underneath his house over the joists that held up the house. Once in a while Dan delivered liquor to the Naples Beach Club Hotel; then it would be stored underwater at an isolated cove, and when needed, a grappling hook was used to retrieve the sacks. Underneath the Methodist Church was another hiding place. A case cost the retailer fifty or sixty dollars—four or five dollars a quart; in turn, the retailer sold each quart, to tourists who could afford it, for eight to ten dollars.

   During his lifetime, Dan often told how he got into rum-running. During the early 1920s he was trout fishing and trying to make a living. He even had 350 acres of tomatoes in the Everglades on a prairie, named Dan House Prairie. The first year he farmed, the rains flooded him out. The next year there was a freeze—you could slice the tomatoes and hear them crack. While visiting a friend on one of the Ten Thousand Islands, a boat came in with sixteen cases of liquor, having been chased by a cutter out of Miami. Dan knew of a man in Everglade who wanted some imported whiskey. He went back to Everglade and arranged to sell the cargo to the man, making fifty dollars per case. Eventually, the demand and profit were so great that he borrowed money and bought a bigger boat that would hold gasoline and food for about two weeks. Coming back from Nassau on one trip, the boat experienced a gasoline explosion and the captain and a Negro helper had to paddle thirty miles back to Bimini in a lifeboat. He would hide his boat in the mangroves—easy to do in the Ten Thousand Islands. He told one of his daughters that he had been searched fifty-two times by the Coast Guard and immigration officers before he was caught. A judge fined him $100, the lawyer cost him $2,000, but his case was dismissed thanks to petitions that, the judge said, made it seem he wasn't such a bad guy after all.

9. George Storter later recounted the tragic death of Hutto, told to him by his friends Gandees and Smith: "The bootleggers, Gene Gandees, Carlton Smith, and a black man, Snoball, figured Christmas Eve would be a good time to bring in a load. They had the sheriff on their payroll, but Hutto wouldn't go along. The chief was tipped off about the shipment and was on patrol about midnight. The chief jumped on the running board, grabbed the windshield stanchion, and the bootlegger's helper reached out, pulled the chief's gun, shot him, and threw the gun after him as he fell in the street and bled to death. The bootleggers were the only witnesses."

10. In 1929 both George and Wesley Storter were arrested and charged with bringing aliens into the United States. Court documents reveal that Wesley was introduced by Miguel Sastre, on or about May 15 of that year in Havana, Cuba, to George Gaci, George Vanda, Vasilie Butnar, and Mihal Thanas, all non-U.S. citizens. Documents further state that on or about June 16, Miguel Sastre and Bernardo Sastre, also known as Antonio Sastre, put eleven aliens aboard the yacht *Delirio* at Havana. The aliens arrived about five miles off the Florida coast in the vicinity of Naples, Florida, on June 17. On June 18, Wesley and Warner Bryant took eleven men aboard another motor vessel—George's boat—and brought them to a point near Naples. George testified that they put them in a small farmhouse in a deserted orange and grapefruit grove.

   The government detained five of the aliens as witnesses, then deported them in December 1929 after they had testified.

   In November, George was found not guilty. The sheriff of Lee County, Frank Tippins, wrote the court in October certifying that to his knowledge George had not violated any law. "I have always considered him to be a good honest hard working boy, and I have known him all of his life." Wesley and Warner were found guilty, sentenced to two years at the Industrial Reformatory at Chillicothe, Ohio, and fined one dollar.

   Rob never wrote about the smuggling incident, but in his brief conversations about it he confirmed that Wes had spent time in prison and it had been a very difficult time for the family. It turns out that Wes borrowed not only George's boat for smuggling, but Rob's as well, at least occasionally. How long this went on, or if Rob knew, is not known.

11. The barometer dropped to 26.35 inches, the lowest sea-level reading in the history of the U.S. Weather Bureau.

12. Although its path was relatively narrow, the 1935 hurricane was one of the most violent on record; the wind reached a velocity of 200 miles an hour, driving a tidal wave over twelve feet high far inland. The hurricane destroyed the railroad between Florida City and Key West. More than five hundred bodies were recovered immediately afterward, and for months unidentified corpses continued to be found in the mangrove swamps. The final number of victims was estimated at eight hundred.

Six. Hunting, Fishing, Nature

1. Ice was packed around the fish to keep them fresh while transporting to the fishhouse instead of their being salted down.

2. Catfish have spined pectoral and dorsal fins that stick straight out and, if caught in the net, won't give; the fisherman has to cut them off or work them out of the mesh, which is likely to tear it.

3. Pelican Key or Comer Island is also known as Bird Key.

4. Federal plume laws enacted in 1900 attempted to save the birds, but they were not successful, as egret plumes could sell for fifty dollars.

Birds were so plentiful and tame it took only a three-foot club to hit a bird and kill it. Between 1905 and 1915 three bird wardens were murdered in the Everglades.

5. Bootleggers used a trail near Whiskey Creek to transport their cargo to the marketplace, and along the creek the remains of old stills and furnaces are still visible.

6. "Hunters," Rob explained, "used a lantern that burned lard and oil. We mixed kerosene and lard together and wore a round lantern strap around our head."

7. Barkley was a Louisiana evangelist who spent a good deal of time in South Florida and Mexico.

8. After Rob moved to Naples, some citizens of Everglade had Sunday morning snake fights at the courthouse—around 1928. They would pit rattlesnakes against king snakes and use the courthouse steps as bleachers.

9. Spring tide, normally the highest tide of the month, occurs at or shortly after the new and the full moon.

10. Deep Lake, Friday Bay, Pinetucky, and Tin Can Hole were other favorite fishing spots.

11. Florence Price's grandfather was Walter Haldeman, founder of the town of Naples.

12. The western novelist Zane Grey, who came to the Everglades around 1924 on a fishing expedition, called it "a country that must be understood."

13. Rob saw Halley's Comet in 1910 when he was sixteen years old. Around 1960 there were three comets with a magnitude of 0 to 1 (as bright as the brightest star in the sky): Arend-Roland and Mykos, both in 1957, and Seki-

Lines, in 1961. All these would have been especially bright on a dark night at Tussok Key.

14. The net, still with fish in it, would be left in the water, marked by a buoy. At any one time fishermen took up only what fish they could carry, leaving the rest to be retrieved later.

15. Rob was so pleased to see bright colors again, especially yellow, which he had not been able to see in a long time.

16. Rob was still fishing alone at age eighty-eight.

17. Rob was ninety-two when he participated in these hog hunts. In one sad respect he was experiencing a new freedom, since Betty had died in February. He had said that his one wish was for her to die before he did—he did not want to leave Marilea alone to care for her.

18. That is, they castrated him, to help keep the boar population down.

Seven. Final Days

1. It was not just that the fish were not there, but Rob said he had gotten "tired of pulling drunks out of the water."

2. When the net ban was passed, Bem moved to northern Florida, though he soon came back home.

3. Named for a coffee mill that was nailed to a tree so everyone in the area could grind their own coffee beans. Coffee was considered a necessity of life.